4D Printing

Additive Manufacturing Materials and Technologies Series

4D Printing

Fundamentals and Applications

Edited by

RUPINDER SINGH

Department of Mechanical Engineering, National Institute of Technical Teachers Training and Research, Chandigarh, India

Series Editor

MA QIAN

ELSEVIER

Elsevier
Radarweg 29, PO Box 211, 1000 AE Amsterdam, Netherlands
The Boulevard, Langford Lane, Kidlington, Oxford OX5 1GB, United Kingdom
50 Hampshire Street, 5th Floor, Cambridge, MA 02139, United States

British Library Cataloguing-in-Publication Data
A catalogue record for this book is available from the British Library

Library of Congress Cataloging-in-Publication Data
A catalog record for this book is available from the Library of Congress

ISBN: 978-0-12-823725-0

For Information on all Elsevier publications
visit our website at https://www.elsevier.com/books-and-journals

Publisher: Matthew Deans
Acquisitions Editor: Christina Gifford
Editorial Project Manager: Joshua Mearns
Production Project Manager: Manju Thirumalaivasan
Cover Designer: Matthew Limbert

Typeset by MPS Limited, Chennai, India

Contents

List of contributors

I.P.S. Ahuja
Department of Mechanical Engineering, Punjabi University, Patiala, India

Ajay Batish
Department of Mechanical Engineering, Thapar Institute of Engineering and Technology, Patiala, India

Kamaljit Singh Boparai
Department of Mechanical Engineering, GZSCCET, MRS Punjab Technical University, Bathinda, India

Abhishek Kumar
Department of Mechanical Engineering, GZSCCET, MRS Punjab Technical University, Bathinda, India

Pawan Kumar
Department of Physics, University Institute of Science, Chandigarh University, Mohali, India

Ranvijay Kumar
Department of Mechanical Engineering, University Center for Research and Development, Chandigarh University, Mohali, India

Sudhir Kumar
Department of Mechanical Engineering, CT University, Ludhiana, India

Vinay Kumar
Department of Production Engineering, Guru Nanak Dev Engineering College, Ludhiana, India; Department of Mechanical Engineering, Punjabi University, Patiala, India

Ravinder Sharma
Department of Mechanical Engineering, Thapar Institute of Engineering and Technology, Patiala, India

Rupinder Singh
Department of Mechanical Engineering, National Institute of Technical Teachers Training and Research, Chandigarh, India

T.P. Singh
Department of Mechanical Engineering, Thapar Institute of Engineering and Technology, Patiala, India

Preface

Thank you for taking the time to read this book on additive manufacturing for 4D applications. I hope that readers will benefit from the time and effort of the authors in preparing this book, which I decided to write after several years of working in the field of additive manufacturing.

Additive manufacturing is currently being incorporated into curricula in many polytechnics and universities around the world. Increasing numbers of students are becoming aware of these technologies and yet, there was not previously a single text adequate for such curricula in 4D applications. I believe that the first edition of this book will provide a text for a clear understanding of the 4D properties of additive manufacturing-based functional prototypes.

I outline the rapid development of this technology from its humble beginnings that showed promise but required much development, to one that is now maturing and showing real benefits to product development organizations. In reading the various chapters, it is hoped that the reader will learn the basics of how additive manufacturing works for 4D applications. The nine chapters (Chapters 1–9) of this book describe in detail the group of additive manufacturing-based 4D applications. First, one has to look at the selection methods for sorting through the many options concerning the type of technology; one should buy in relation to the application and so guidelines are provided on how to select the right technology for a particular purpose. One may now consider the possibility of mass customization, where a product can be produced according to the requirements of an individual consumer but at a cost-effective price. This moves us on nicely to the subject of the applications of additive manufacturing, including tooling and products in the medical, manufacturing, and construction industries.

This book is primarily aimed at students and educators studying additive manufacturing for 4D applications, either as a self-contained course or as a module within a larger course on manufacturing technology. There is sufficient depth herein for an undergraduate or graduate-level course, with many references that can provide the reader with greater detail and specifics. Additive manufacturing researchers may also find this text useful in helping them to understand the state of the art and the opportunities for further research.

Although I have worked hard to make this book as comprehensive as possible, I recognize that a book about such a rapidly changing technology will not be up-to-date for very long. With this in mind, and to help educators and students better utilize this book, I will update our course website, with additional homework exercises and other aids for educators.

If you have comments, questions, or suggestions for improvement, they are welcome. I anticipate updating this book in the future, and look forward to hearing how you have used these materials and how to improve this book.

As mentioned earlier, each author is an established expert in additive manufacturing with a number of years of research experience. In addition, this book is only possible thanks to the many students and colleagues with whom I have collaborated over the years.

Rupinder Singh
Department of Mechanical Engineering, National Institute of Technical Teachers Training and Research, Chandigarh, India

Additive manufacturing for 4D applications

1 Introduction

The book *Additive Manufacturing for 4D Applications* explores both autonomic and nonautonomous systems with a different stimulus such as temperature, current, moisture, light, and sound. In addition, the fourth-dimensional aspect using more than one stimulus is outlined for additive manufacturing processes. It presents both an introduction to the basic understanding of hybrid processes, as well as exploring the physics behind the process. For the field engineers, applicable codes and standards for each hybrid process are provided. Lastly, case studies are included in each chapter and will provide the reader with a model to explore future research directions. This book focuses on hybridization of various established additive manufacturing techniques along with the use of smart consumables for possible 4D applications. Overall this book follows following strategy:

- Begins with the fundamentals of the hybrid additive manufacturing process
- Presents a discussion of the physics behind the smart material functioning in hybrid additive manufacturing
- Includes real-world case studies on 4D printing, as well as a look at the future research dimensions in 4D printing

In order to ascertain the research potential, Scopus data (Database source: www.scopus.com) has been explored for the keyword "additive manufacturing for 4D applications" and 177 documents were noticed from 2010 to 2021. Further VOS viewer open source software was used for the analysis of these documents and by keeping the number of keywords "5," out of 1465, 82 met the threshold. For the selected 82 keywords, the total strength of cooccurrence link with other keywords was calculated (Table 1). Based upon Table 1, Fig. 1 shows bibliographic analysis for keyword "additive manufacturing for 4D applications."

Based on Fig. 1, Figs. 2–4 shows the research gap in the proposed area of 4D applications using additive manufacturing.

Based upon these selected documents, Figs. 5–8, respectively, show relevant publications in the concerned area, bifurcation based upon a subject area, document type, documents by country or territory in past one decade on additive manufacturing for 4D applications.

Table 1 Total strength of cooccurrence for keyword "additive manufacturing for 4D applications."

Id	Keyword	Occurrences	Total link strength
1.	3D printing	33	305
2.	3D printers	96	710
3.	3D printing	57	411
4.	3D printing process	5	32
5.	4D printing	101	644
6.	Additive manufacturing	80	535
7.	Additive manufacturing (am)	6	41
8.	Additive manufacturing process	9	81
9.	Additive manufacturing technology	12	93
10.	Additives	44	302
11.	Biocompatibility	10	90
12.	Biodegradable polymers	5	45
13.	Biomaterial	5	52
14.	Biomaterials	10	87
15.	Biomedical applications	5	48
16.	Biomimetics	9	46
17.	Bioprinting	14	142
18.	Chemistry	6	67
19.	Complex geometries	5	38
20.	Composite materials	6	61
21.	Composite structures	8	55
22.	Computer aided design	11	92
23.	D-printing	50	410
24.	Deformation	7	40
25.	Design/methodology/approach	6	46
26.	Digital light processing	5	32
27.	Drug delivery system	5	49
28.	Energy harvesting	5	36
29.	External stimulus	7	58
30.	Extrusion	6	52
31.	Fabrication	14	118
32.	Flexible electronics	8	57
33.	Functional materials	6	51
34.	Functional polymers	8	68
35.	Fused deposition modeling	13	89
36.	Future perspectives	6	52
37.	Geometry	5	29
38.	Human	9	102
39.	Humans	9	102
40.	Hydrogels	5	43

(Continued)

Table 1 (Continued)

Id	Keyword	Occurrences	Total link strength
41.	Industrial research	8	56
42.	Inkjet printing	5	34
43.	Intelligent materials	20	173
44.	Manufacture	19	167
45.	Manufacturing techniques	12	96
46.	Manufacturing technologies	6	41
47.	Medical applications	6	56
48.	Multi materials	6	53
49.	Origami	6	42
50.	Photopolymerization	7	42
51.	Polymer	5	35
52.	Polymers	20	159
53.	Printable materials	6	53
54.	Printed structures	8	54
55.	Printing	21	184
56.	Printing presses	8	57
57.	Printing, three-dimensional	11	121
58.	Procedures	5	62
59.	Product design	10	73
60.	Rapid prototyping	8	56
61.	Recovery	6	51
62.	Reinforced plastics	5	22
63.	Review	5	41
64.	Robotics	6	39
65.	Scaffolds (biology)	9	85
66.	Selective laser melting	5	41
67.	Shape-memory effect	20	178
68.	Shape-memory polymer	6	60
69.	Shape-memory polymers	21	158
70.	Shape optimization	6	67
71.	Shape-memory materials	11	98
72.	Shape-memory polymer	19	140
73.	Shape-memory properties	5	36
74.	Smart materials	25	169
75.	Soft robotics	7	51
76.	State of the art	5	33
77.	Stereolithography	8	70
78.	Stimuli-responsive	8	61
79.	Three-dimensional printing	12	122
80.	Three-dimensional printing	12	100
81.	Tissue	8	91
82.	Tissue engineering	12	116

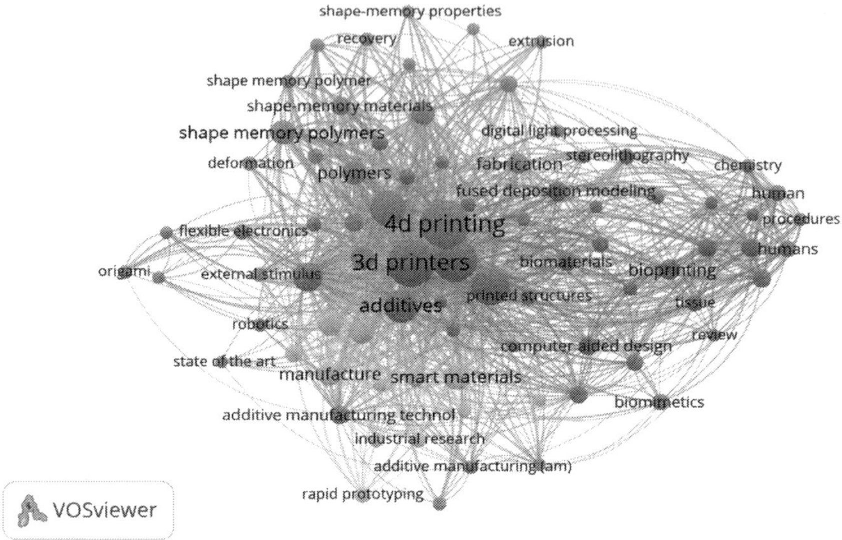

Figure 1 Bibliographic analysis for keyword "additive manufacturing for 4D applications."

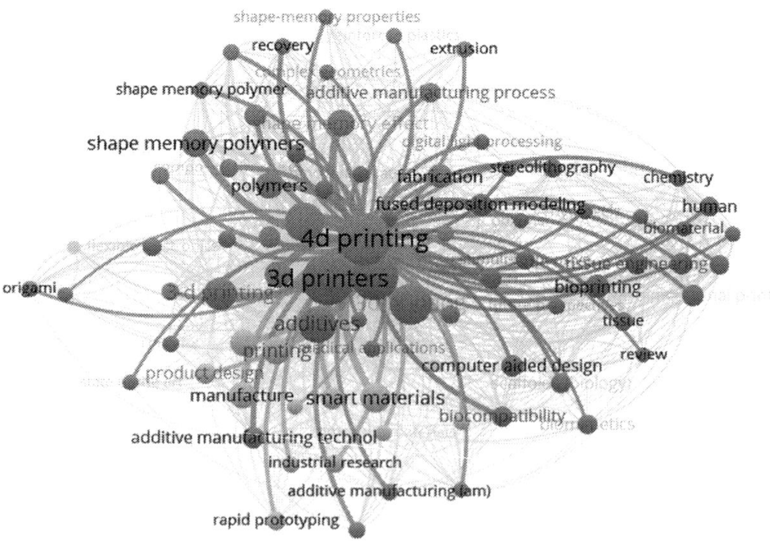

Figure 2 Research gap analysis with node as 4D printing.

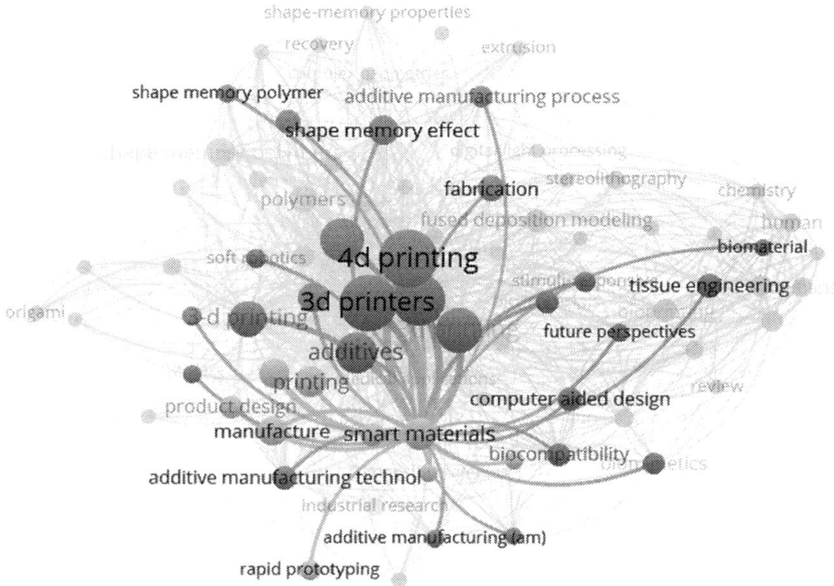

Figure 3 Research gap analysis with node as smart materials.

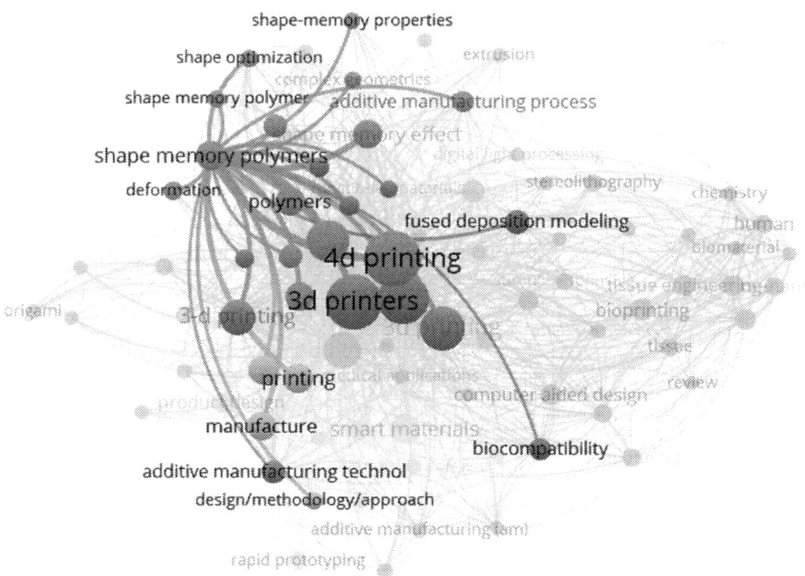

Figure 4 Research gap analysis with node as shape-memory polymers.

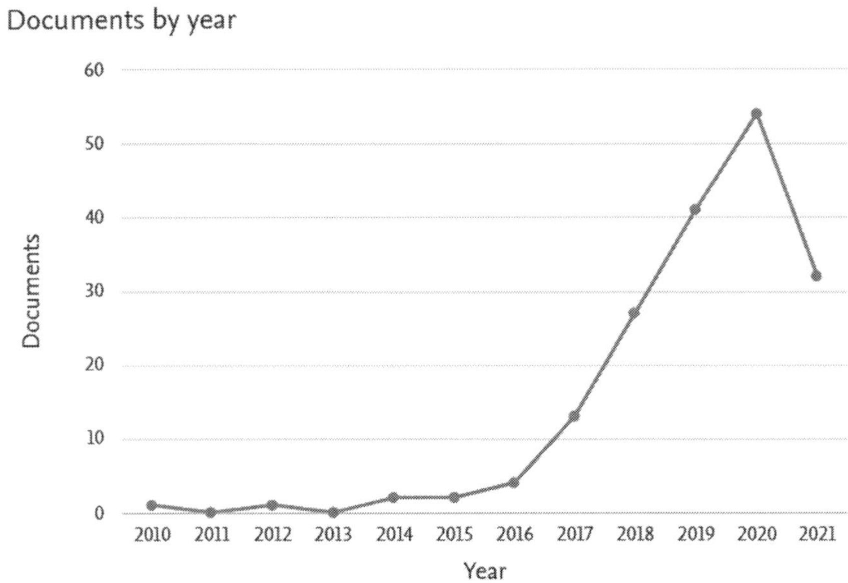

Figure 5 Research publication in past one decade for additive manufacturing for 4D applications.

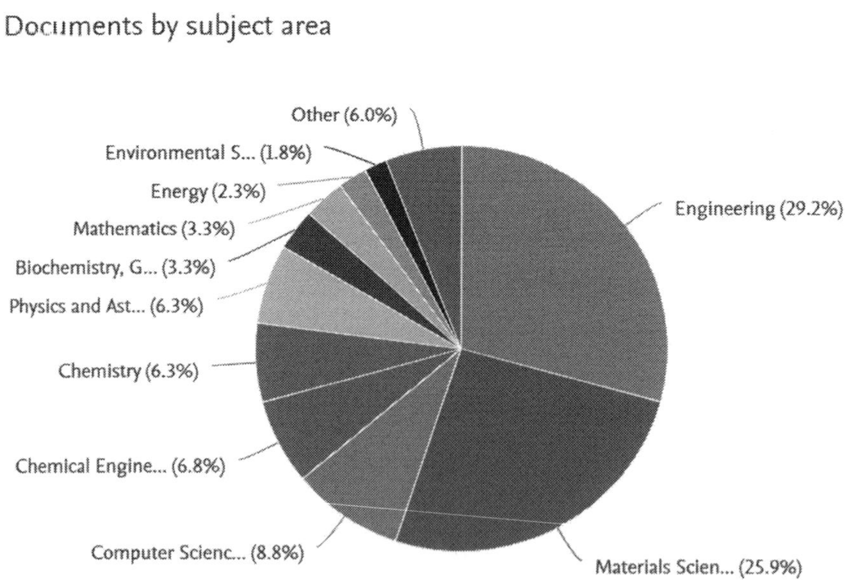

Figure 6 Research publication in past one decade on different subject areas for additive manufacturing for 4D applications.

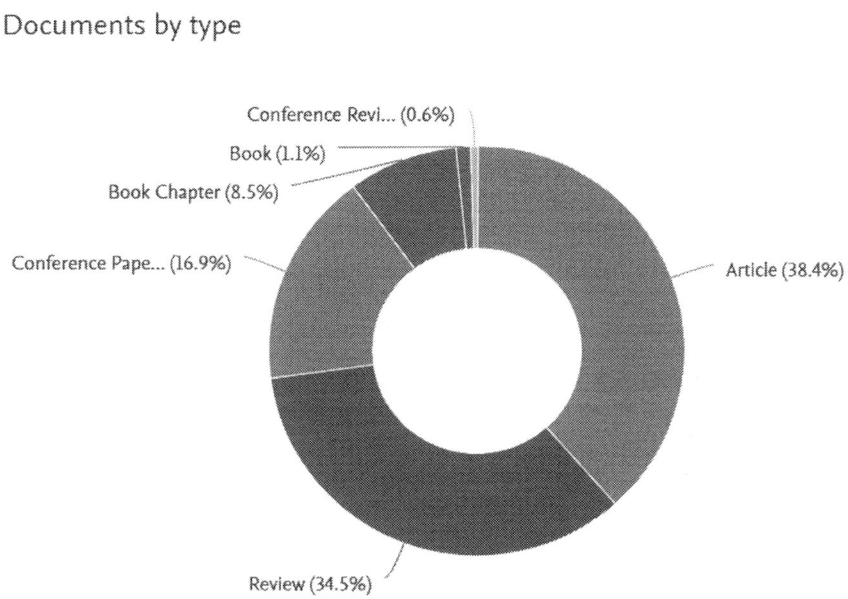

Figure 7 Research publication in past one decade based upon the type of document published on additive manufacturing for 4D applications.

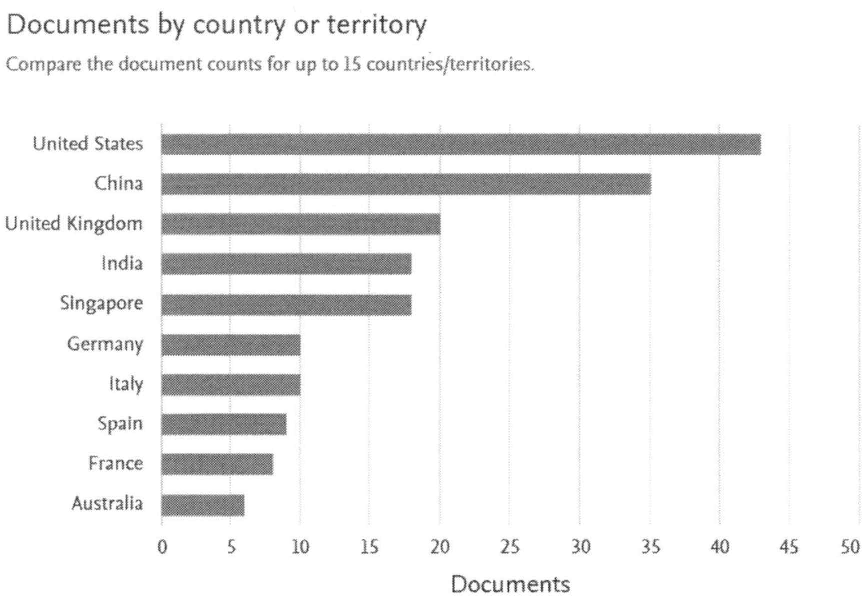

Figure 8 Research publication in past one decade at the international level on additive manufacturing for 4D applications.

This book comprises of nine chapters. The chapter highlights are given in the following:

Chapter 1: On 3D Printed Multi blended and Hybrid-Blended Poly(lactic) Acid Composite Matrix for Self-Assembly

This chapter highlights the investigation on hybrid and multimaterial matrix of polylactic acid (PLA) composite prepared by using wood powder, polyvinyl chloride, and Fe_3O_4 powder as reinforcement. The functional prototypes of different compositions were explored for 4D applications. The case study result gave a clear indication of feasibility for the multimaterial and hybrid-blended functional prototype 3D printing. Vibration sample magnetometry (VSM)—based analysis has shown that the magnetization of the multimaterial and hybrid material-based composite was similar, which meant that both of the composites can show self-assembly characteristics on the application of an external magnetic field.

PLA PLA/Wood powder PLA/PVC

PLA/Fe_3O_4 powder

Graphical highlights: Process flow for hybrid material matrix.

Chapter 2: Graphene-Reinforced Acrylonitrile Butadiene Styrene Composite as Smart Material for 4D Applications

The present study reported the chemical blending of acrylonitrile butadiene styrene (ABS) and graphene (G) to prepare a smart composite material with good magnetic and electric conductivity properties along with a significant shape-memory effect. The blends of G reinforced ABS (GABS) composite were obtained by changing the weight proportion of G from 0% to 20%. Each composition/proportion of GABS was twin-screw extruder (TSE) to prepare wire specimens for investigating shape-memory effect and VSM-based magnetic properties.

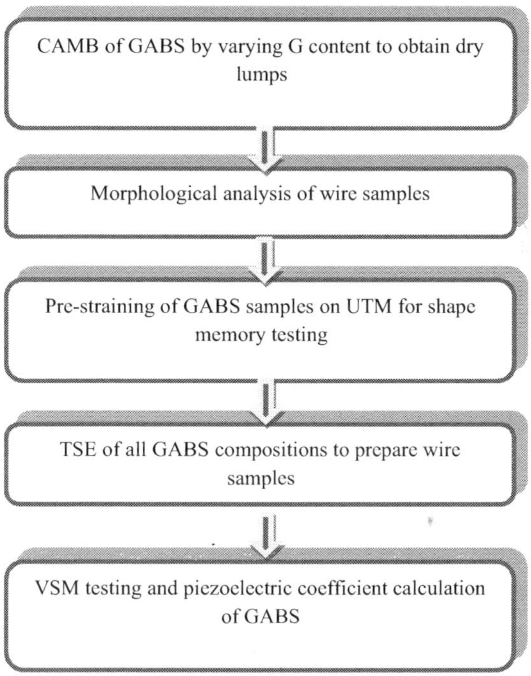

Graphical highlights: Work methodology for 4D characterization of GABS composite.

Chapter 3: Two-Way Programming Of Secondary Recycled Poly (lactic) Acid Composite Matrix Using Magnetic Field as Stimulus

In this study, feedstock filament of recycled PLA composite reinforced with Fe_3O_4 powder was developed on lab-scale for three recycling stages and the developed feedstock was tested for mechanical, magnetic, and morphological characteristics. In the end, the magnetization capability of the developed feedstock was tested for its repeatability and reliability by calculating its performance statistically over the different recycling stages of polymeric composite.

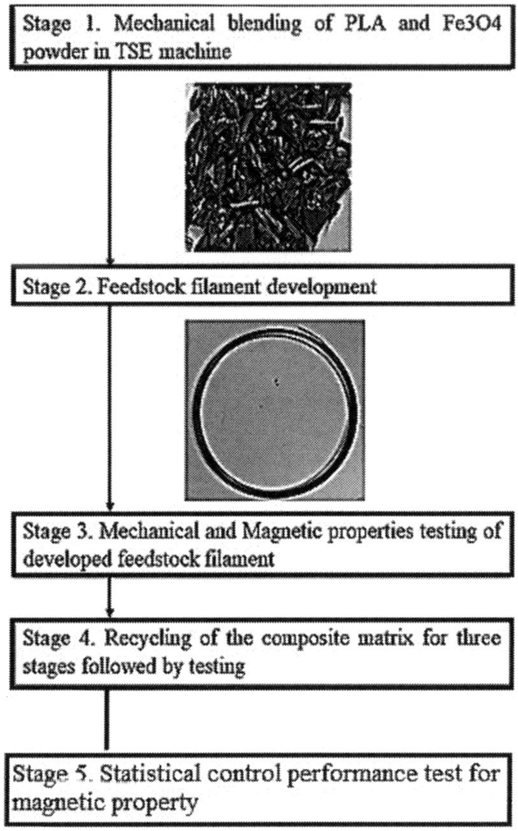

Graphical highlights: Methodology followed in the present case study.

Chapter 4: 3D Printed Graphene Reinforced Polyvinylidene Fluoride Composite for Piezoelectric Properties

This study reports the investigations performed on the composite of polyvinylidene fluoride (PVDF) in which graphene nanoparticles were blended with chemical assistance to fabricate 3D printed prototype and testing the piezoelectric and magnetic properties for in-house developed composites. In the present study, three different compositions/proportions of PVDF and PVDF-graphene (PVDFG) composites were prepared. Chemical-assisted mechanical blended (CAMB) of PVDFG composites were obtained by blending 5% (PVDFG5) and 10% (PVDFG10) graphene by weight proportion in PVDF. The PVDF, PVDFG5, and PVDFG10 composites prepared by CAMB were processed on TSE to fabricate 3D printer feedstock filament wire. The 3D printed PVDFG composites were then tested for piezoelectric coefficient or piezoelectric modulus (D_{33}) and VSM-based magnetic properties for characterization of 4D behavior.

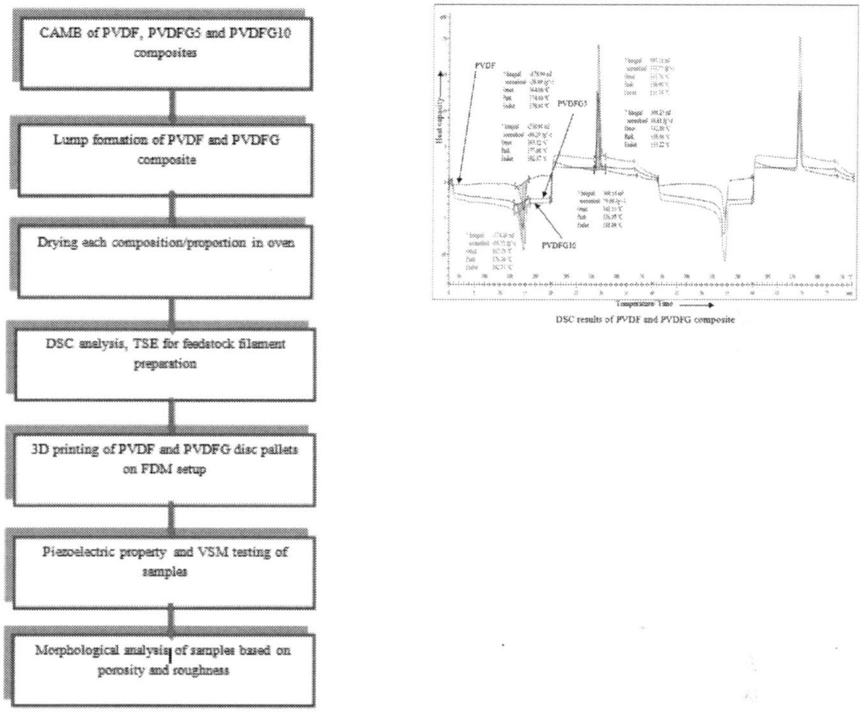

Graphical highlights: Work methodology for CAMB of PVDF and PVDFG composites 4D analysis.

Chapter 5: On Characterization of Rechargeable, Flexible Electrochemical Energy Storage Device

In this study paraffin wax in varying proportion was mechanically blended within graphite and graphene. The new stable compositions/proportions were developed and extruded via TSE in order to fabricate feedstock filament. Structural analysis was visualized by using scanning electron microscopy and Fourier transform infrared spectroscopy (FT-IR). Melt flow index (MFI) testing was carried out to understand the rheological behavior of newly developed novel materials while the thermal properties have been measured and by using differential scanning calorimetry to explore 4D capabilities.

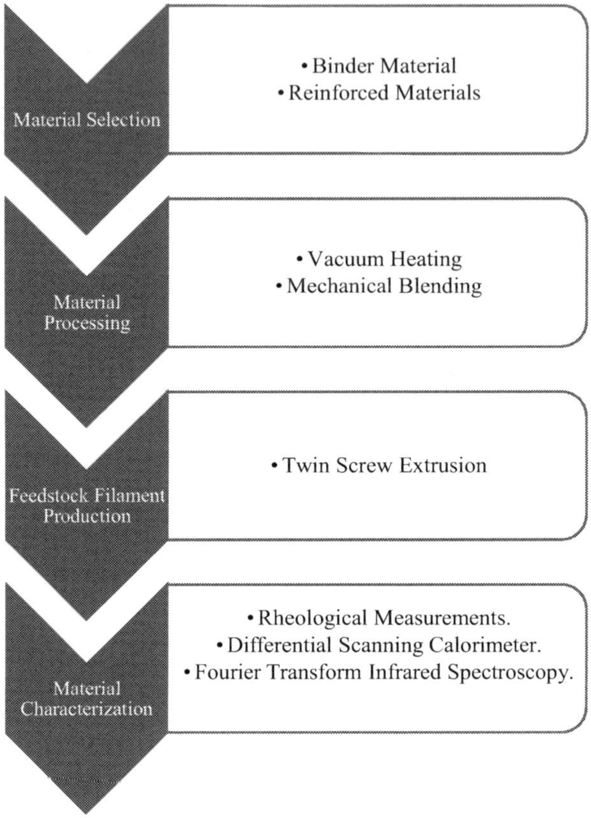

Graphical highlights: Process flow diagram.

Chapter 6: On Dual/Multi material Composite Matrix for Smart Structures: A Case Study of ABS-PLA, HIPS-PLA-ABS

In this study an effort has been made to analyze different material combinations for 3D printing of functionally graded multimaterial prototypes. This work is an extension of previously reported work on 3D printed prototypes of dual material (PLA and ABS) in which it was suggested that the material combination must be properly selected for better mechanical performance and a thumb rule was derived which considers the number of conversions and number of negative conversions based on glass transition temperature (T_g) of material for better mechanical results.

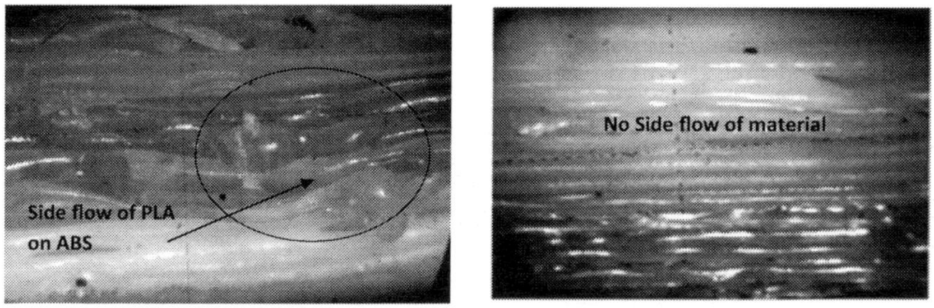

(Base: ABS; Middle layer: PLA; Top Layer: HIPS) (Base layer: PLA; Middle layer: HIPS; Top Layer: ABS)

(a)

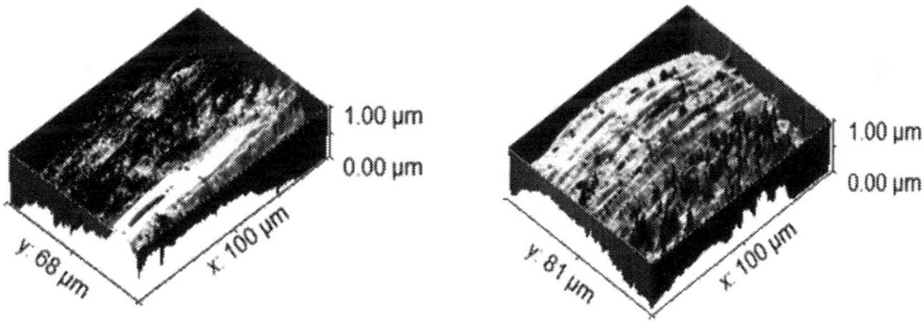

(b)

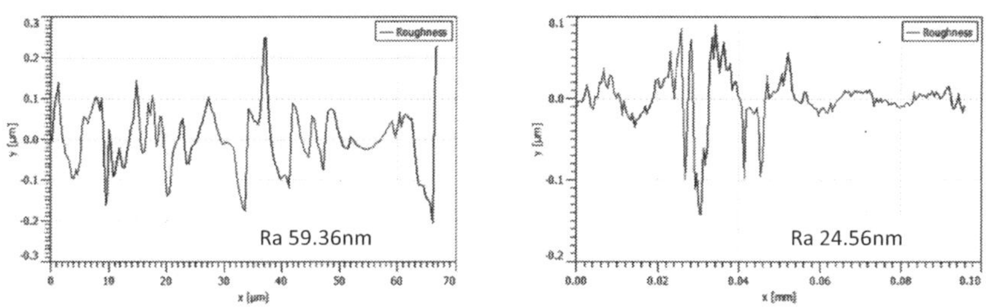

(c)

Graphical highlights: (a) Tool maker microscopic images for samples, (b) 3D rendered image for samples, and (c) surface roughness profile for samples.

Chapter 7: PVDF-Graphene-BaTiO3 Composite for 4D Applications

This study presents the mechanical, dimensional, and 4D properties of PVDF/Graphene (Gr) /$BaTiO_3$ composites developed via chemical mixing method. The PVDF was selected as a base polymer matrix due to its flexible, light-weight, and high piezoelectric properties. As per the previously reported studies on mechanical blending, the proportion of Gr and

$BaTiO_3$ was 2% and 15% by weight in polymer matrix. Standard tensile specimens as per ASTM D-638 were 3D printed using a feedstock filament prepared from a developed composite.

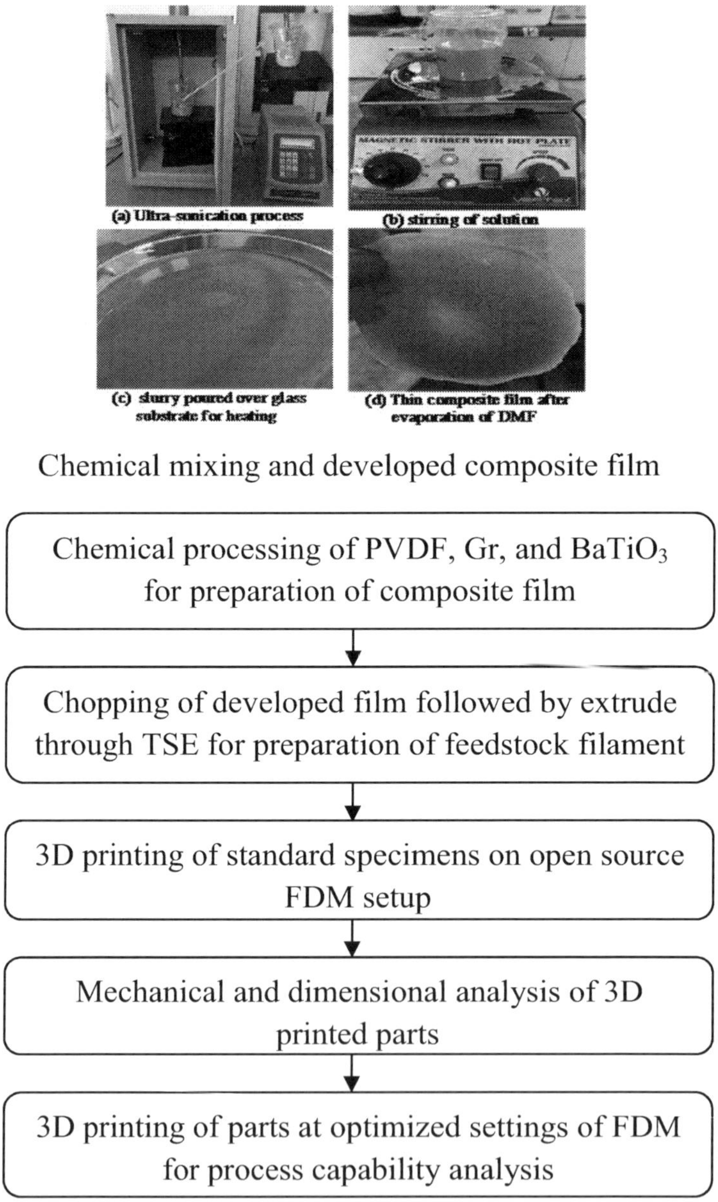

Graphical highlights: Methodology adopted for this research work.

Chapter 8: Hydrothermal Stimulus for 4D Capabilities of PA6-Al-Al2O3 Composite

In this chapter an experimental investigation has been reported to realize the 4D capabilities of polyamide (PA)6-Al-Al_2O_3 based composite feedstock filament for fused deposition modeling process. The change in length of the composite feedstock filament specimens have been tested by stretching on tensile testing machine with a rate of 30 mm/min for 10 seconds, and after releasing the load, the specimens have been exposed to room temperature conditions for 2 hours. The four types of testing conditions were considered, that is, specimens without any stimulus, specimens with water, heat, and hydrothermal stimulus.

Graphical highlights: MFI testing as per ASTM D1238 and tensile testing as per ASTM D-638.

Chapter 9: On PLA–ZnO Composite Matrix for Shape Memory Effect

In this study, the PLA feedstock filaments with zinc oxide (ZnO) nanoparticles reinforcement have been manufactured by using twin-screw compounding process. The PLA-ZnO composites in feedstock filaments form have been manufactured for the FFF process by considering the preheat and post heat treatment with varying processing temperature and screw torque. The shape-memory investigations of manufactured PLA-ZnO composite filaments have been performed by taking temperature as the external stimulus.

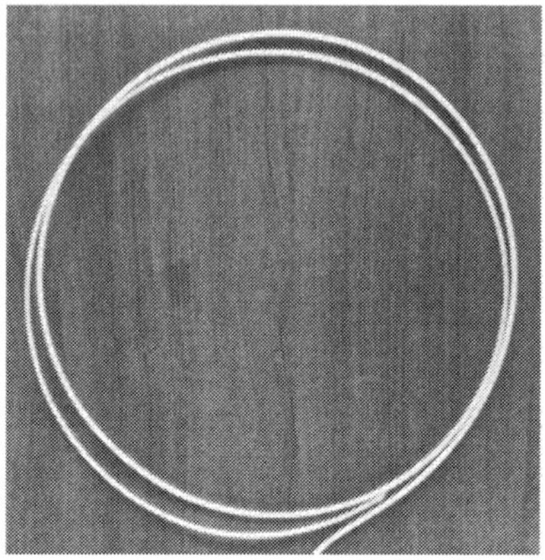

Graphical highlights: Twin-screw compounder and feedstock filament.

This book is extensively illustrated to convey a thorough understanding of the additive manufacturing processes for 4D applications and the equipment involved. Just a few figures have been purposely repeated to avoid a break in thoughts during reading to search them elsewhere in the book. It is expected that the book will be useful for not only the bachelors/ masters students and research scholars but also for the practicing engineers in this field.

In spite of the best of my efforts, there are bound to be some mistakes. The same may kindly be brought to my notice for rectification. Any other suggestions to improve the book will be thankfully acknowledged.

I sincerely wish that the book may come up to the expectations of the readers on this vital subject of additive manufacturing technology.

Rupinder Singh

CHAPTER 1

On 3D printed multiblended and hybrid-blended poly(lactic)acid composite matrix for self-assembly

Sudhir Kumar[1], Rupinder Singh[2], T.P. Singh[3] and Ajay Batish[3]
[1]Department of Mechanical Engineering, CT University, Ludhiana, India
[2]Department of Mechanical Engineering, National Institute of Technical Teachers Training and Research, Chandigarh, India
[3]Department of Mechanical Engineering, Thapar Institute of Engineering and Technology, Patiala, India

1.1 Introduction

3D printing of polymeric material started in the late 1980s and the very first technology to start the material printing was stereo-lithography and after that there was a series of advancement in the field of functional prototype 3D printing using different ways [1,2]. Over the period of time, the 3D printing has slowly evolved in 4D printing of objects. The 4D refers to the change in material characteristics with "time" [3,4]. Various studies have proved that the 4D nature of printing comes from the ability of 3D printed object to change its shape/geometry when triggered externally or internally. Thus the 4D behavior is a combination of material, geometry, and functionality [5–8].

The difference in 3D and 4D lies in the fact that the structure of 3D printed specimens is generally static and has no changes in structure over a period of time [9,10], while in 4D printing, the printed object contains the smart material matrix that is printed with the smart geometric structure (mathematically calculated) and has at least one external or internal trigger to stimulate the required change [11–13]. The researchers have used several external stimuli for stimulating the change in the shape characteristics of the material. Some of the external stimulus being used by researchers are water [14,15], heat [5,16], and water and heat together [17,18].

The most critical parameter of 4D printing is the material matrix. The material may have properties that are essential to 4D behavior such as

4D Printing
DOI: https://doi.org/10.1016/B978-0-12-823725-0.00003-5

expansion and contraction when some changes in external environment take place [19]. Some material matrixes may be made stimulus responsive by adding some foreign reinforcement in the base material matrix [18,20]. Some of the properties that have greater significance in 4D world are self-sensing, self-healing, self-assembling, shape memory responsiveness, decision making, etc. [21–25].

In recent times, researchers have worked on the development of smart material matrixes which can easily respond to the external stimulus already selected on basis of reinforcement in base material matrix such as magnetic field for magnetite powder, iron (Fe) powder-based reinforcement, similarly water for hydrogel-based materials, etc. [26–29]. Some of the research groups have done extensive work for the development of range of smart material matrixes by incorporating diverse range of novel particles of various sizes such as carbon nanotube provides extra flexibility and strength to material matrix, glass fiber gives high impact resistance, nano-size particle of silicon carbide introduces extra surface roughness, and aluminum oxide (Al_2O_3) provides low wear rate.

Similarly, various other materials are available these days in market which could be added to polymeric base material and can provide the required characteristics and can enhance the mechanical, thermal, electrical, and other properties of base material matrix [30–32]. The only purpose of introducing the foreign reinforcement in the base polymeric matrix is to improve the material property and make it effective for 4D transitions such as extra flexibility will provide bending or folding capability to 3D printed material [33–38]. The capability of fused deposition modeling (FDM) machine can also provide some improvement in mechanical performance of 3D printed objects such as input process parameters (infill density, infill pattern, infill ratio, infill angle, raster angle, raster width, etc.) have great role to play in properties of 3D printed object [39–42].

From literature review, it has been found that the 4D behavior largely depends over the material matrix and part geometry to be printed. But 3D printing of multimaterial and hybrids blend has not been fully explored for self-assembly application. Therefore, in this study, a case study has been performed to develop hybrid blend and multimaterial matrix of poly(lactic) acid (PLA) for 3D printing functional prototypes and the results were compared to know which material matrix and geometry of 3D printed object is superior by testing those 3D printed object for mechanical (tensile and flexural) and morphological properties [scanning electron microscopy (SEM), Fourier transformation infrared radiation (FTIR)].

1.2 3D printing of hybrid and multiblended matrix of poly (lactic)acid: a case study

1.2.1 Preparation of hybrid-blended matrix and its printing on fused deposition modeling platform

For 3D printing of hybrid material matrix of PLA on FDM platform, mechanical blending of PLA, polyvinyl chloride (PVC), wood powder and magnetite powder (Fe_3O_4) was performed. PLA and PVC polymer were precured from local market (Chemicals for Lab, Ludhiana, India). Magnetite powder was purchased from Shiva Chemicals, Ludhiana of 44 micron size. The wood powder from Different trials were made to select the proper material matrix ratio for hybrid blending using melt flow index (MFI) testing and it was found that PLA: 50 wt.%, PVC: 25 wt.%, wood powder: 5 wt.%, and Fe_3O_4 powder: 20 wt.% were the optimum material matrix ratio in which there was no difficulty in flow of material matrix through MFI tester thus the selected material matrix ratio would also easily pass through FDM nozzle without causing any difficulty such as choking of die. The developed PLA composite matrix was then processed for feedstock filament development using extrusion machine (Twin Screw Extrusion; Model: HAAKE, Germany) for development of in-house feedstock filament for FDM printer and finally those developed feedstocks were processed for 3D printing of tensile and flexural specimens. Fig. 1.1 shows the process flow of hybrid material matrix of PLA and real-time (A) polymer and hybrid-blended matrix, (B) feedstock and printed functional prototypes for (C) tensile specimen according to ASTM D638 type IV, and (D) flexural specimens according to ASTM D790.

1.3 Preparation of multimaterial matrix of poly(lactic)acid and its printing on fused deposition modeling platform

The polymer used for preparation of multimaterial matrix of PLA was same as that of hybrid blend material matrix but this time four different material matrices were prepared rather than single. The four different material matrix composite ratios were again selected on the basis of MFI which was in the range of 37–42 (g/10 min) for (1) PLA (2) PLA in reinforcement with PVC (75/25 wt.%), (3) PLA in reinforcement with Fe_3O_4 powder (80/20 wt.%) except for the PLA/wood powder (95/5 wt.%) whose MFI was in range of 27–29 (g/10 min). After preparation of material matrices, similar procedure of 3D printing the functional

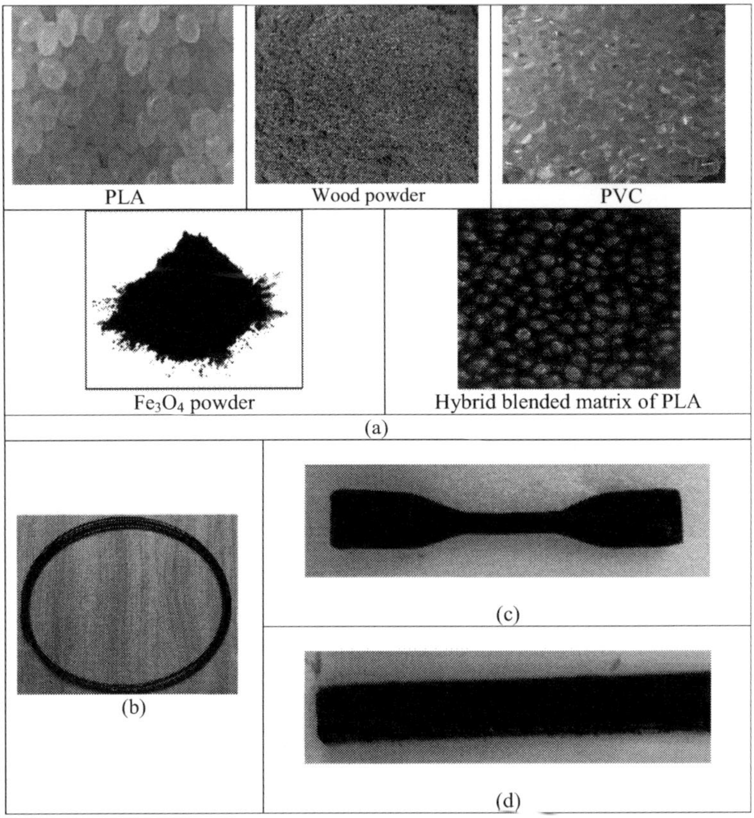

Figure 1.1 Process flow of for hybrid material matrix (A) different material matrix and hybrid material blend, (B) developed feedstock and functional prototypes for (C) tensile specimen according to ASTM D638 type IV, and (D) flexural specimens according to ASTM D790.

prototypes was used as it was used in case of hybrid material matrix. Fig. 1.2 shows the process flow for multimaterial matrix of PLA (A) different polymeric composites, 3D printed (B) tensile specimen (top view), (C) tensile specimen (side view), (D) flexural specimen (top view), and (E) flexural specimen (side view).

1.4 Comparative results of multimaterial and hybrid-blended matrix of poly(lactic)acid

The 3D printing of samples was performed using constant condition of printing for different functional prototypes such as extrusion temperature 210°C, speed of infill 90 mm/min, and infill pattern rectangular and

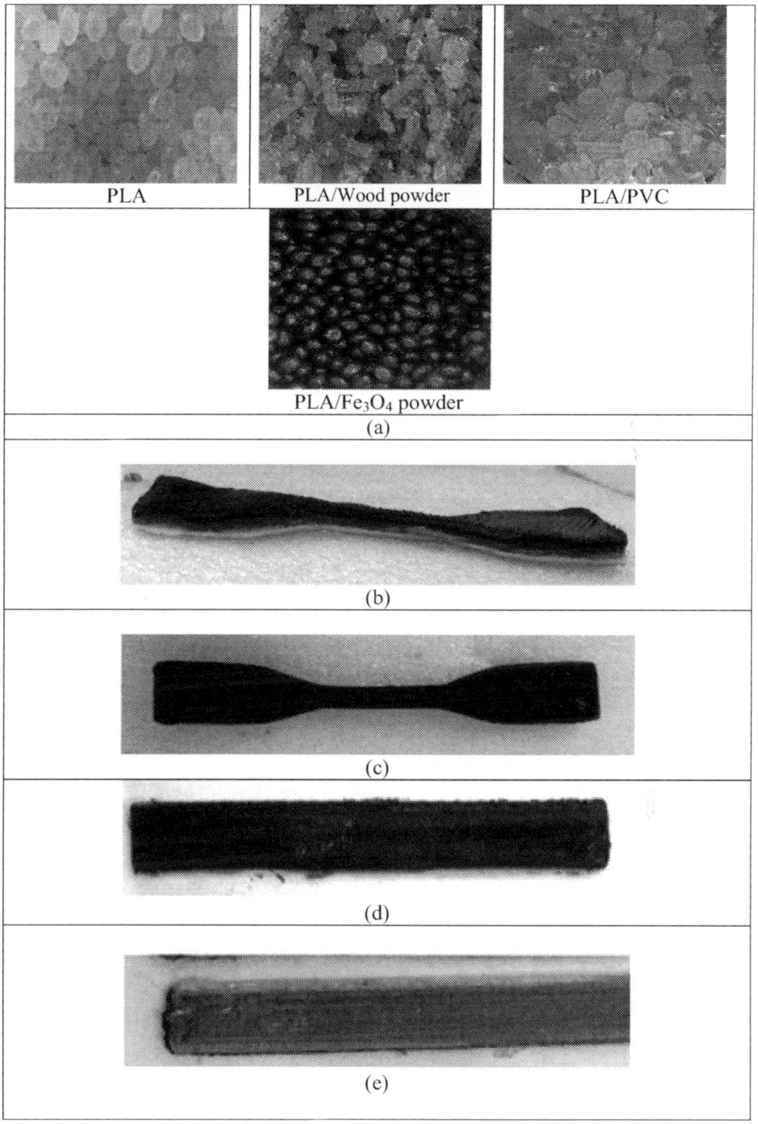

Figure 1.2 Process flow for hybrid material matrix (A) different material matrix of multimaterial blend, 3D printed (B) tensile specimen (front view), (C) tensile specimen (top view), (D) flexural specimen (top view), and (E) flexural specimen (bottom view).

constant infill speed of 100 mm/min. The developed 3D printed tensile and flexural specimens of hybrid blend-based PLA composites and multimaterial-based composites were tested with universal tensile machine

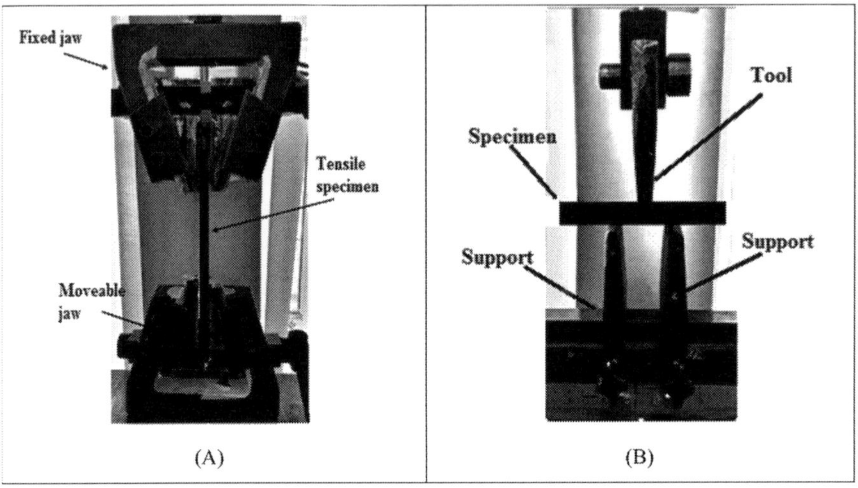

Figure 1.3 UTM setup for (A) tensile testing and (B) flexural testing.

(UTM) with similar testing condition such as testing speed of 30 mm/min. Fig. 1.3 shows the tensile and flexural testing setup of UTM machine. The obtained results for hybrid material matrix and multimaterial matrix were compared for best type of sample so that the material matrix with best mechanical performance.

1.5 Mechanical properties of multimaterial and hybrid-blended matrix

1.5.1 Tensile and flexural properties

Table 1.1 shows the different tensile properties of hybrid-blended and multimaterial matrixes of PLA for different types of samples. From Table 1.1, it has been figured out that the multimaterial matrix-based tensile and flexural samples are superior to those hybrid-blended material-based 3D printed prototypes. This may be due to the fact that the multimaterial matrix-based 3D printed samples held large wt.% of PLA in comparison with hybrid material blend. The different layers of material matrix in multimaterial matrix have shown proper fusion of transition layers as there were three transitions in which the base layer was of PLA (100 wt.%), one layer of PLA/PVC, two layers of PLA/wood powder, and at the end last two layers were of PLA/Fe_3O_4 powder. Thus the multimaterial matrix has shown proper compatibility toward 3D printing of different material in successive layers which resulted into better property

Table 1.1 Mechanical performances of different functional prototypes of multimaterial and hybrid-based 3D printed functional prototypes of PLA.

Sample type	PS (MPa)	BS (MPa)	MoT (MPa)
Tensile hybrid material matrix	29.5	25.6	1.62
Tensile multimaterial matrix	46.5	42.0	4.76
Flexural hybrid material matrix	14.85	12.34	2.82
Flexural multimaterial matrix	26.90	24.30	5.57

BS, Break strength; *MoT*, modulus of toughness; *PS*, peak strength.

and has shown result at par with PLA-based material. Whereas in case of hybrid material matrix, single material matrix has resulted into poor results which may be due to the fact that single material matrix may have several problems inside the 3D printed object such as porosity/voids, surface roughness, etc. The single material matrix on the contrary was easy to 3D print as there was no human interaction once the feedstock filament was loaded in FDM machine but in case of multimaterial matrix-based specimen human interaction was necessary to change the feedstock at every stage wherever there was change in material layer while 3D printing.

Fig. 1.4A–D shows the stress vs. strain diagram for different material matrixes and functional prototypes. From Fig. 1.4A–D, it was observed that the stress and strain capacities of multimaterial matrix were high in comparison with those with hybrid blend-based matrix of PLA.

1.6 Morphological properties

1.6.1 Scanning electron microscopy analysis

The SEM characteristics of multimaterial and hybrid material matrix were necessary to explore as the properties obtained for PLA composite matrix differed greatly (near 150% less mechanical performance of hybrid blend-based PLA). The samples made on 3D printer were subjected to UTM testing first, and the tested samples were used to cut sections of 3D printed sample and observed under SEM. It should be noted that for multimaterial sample different sections of layers were observed as there were three transition layers of PLA/PVC, PLA/wood powder, and PLA/Fe_3O_4 powder. From SEM analysis of samples, it was observed that the Fe_3O_4 particles in case of hybrid material matrix moved inside the material matrix of 3D printed layers and the surface obtained was very rough. The SEM analysis of internal section of samples has made it clear that the hybrid

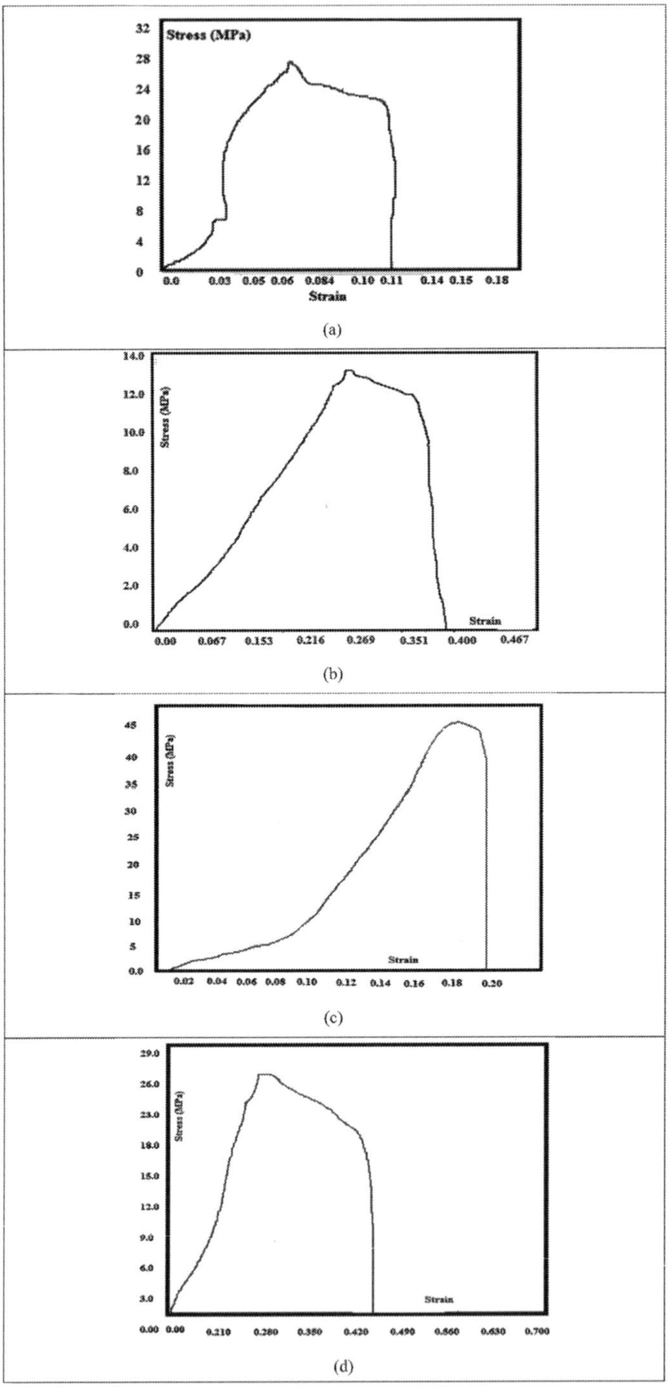

Figure 1.4 Stress versus strain plots for (A) tensile specimen of hybrid material matrix, (B) flexural specimen of hybrid material matrix, (C) tensile specimen of multimaterial matrix, (D) flexural specimen of multimaterial matrix.

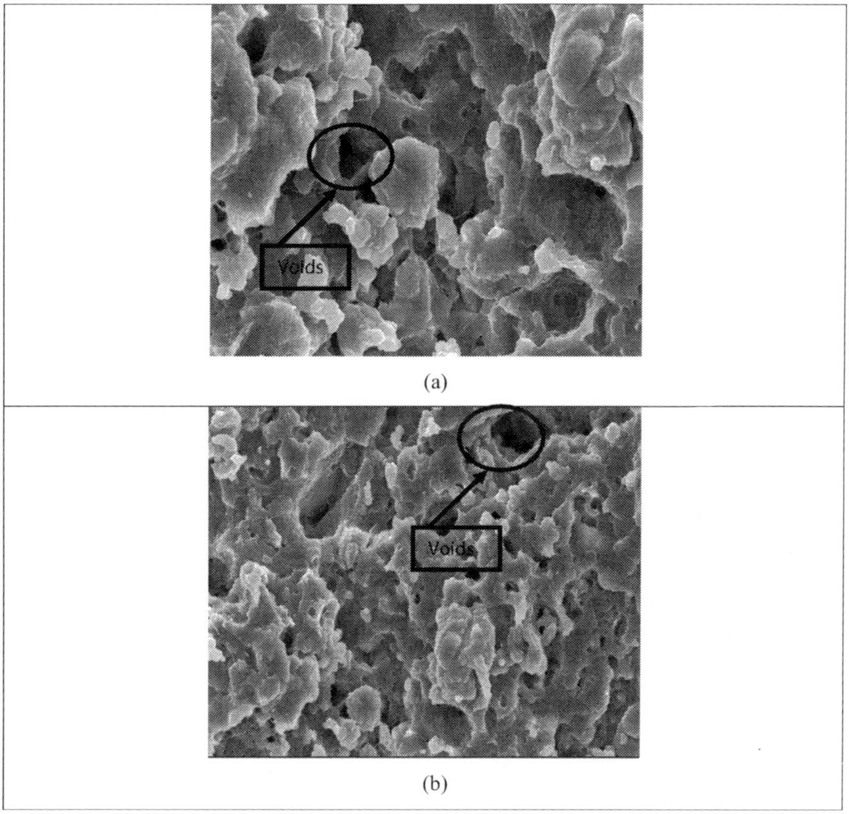

Figure 1.5 SEM images for hybrid material matrix for (A) tensile and (B) flexural specimen.

material matrix has several porosity holes inside the 3D printed geometry (Fig. 1.5).

The SEM analysis of different sections of multimaterial matrix has shown that the transition layers have proper fusion of carbon element (as the SEM suggested see Fig. 1.6). Moreover, the layers had less surface roughness in comparison with hybrid blend-based material.

1.6.2 Fourier transformation infrared spectroscopy analysis

It is well established that the different functional groups of PLA can be found in 1100−4000 range of wave number (WN) such as C=O: 1720, -C-O: 1100−1220, and −CH3: 1450 [43]. Table 1.2 shows the FTIR absorbance spectrum magnitude for different groups for different PLA composite. The different PLA composites have shown increase in the

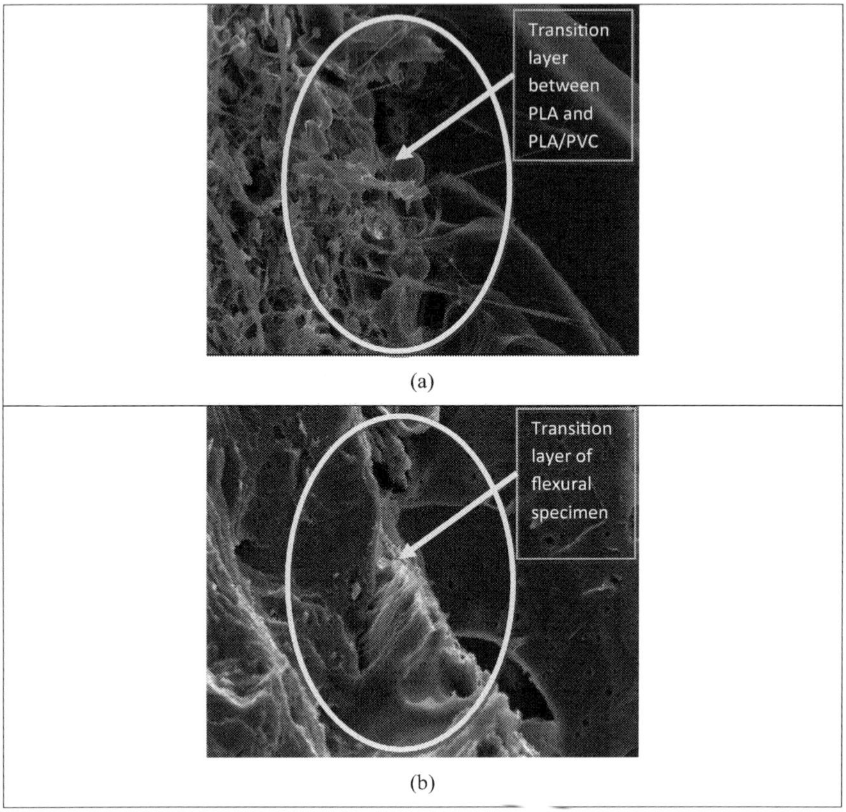

Figure 1.6 SEM images for hybrid material matrix for (A) tensile and (B) flexural specimen.

Table 1.2 Percentage age absorbance for different functional groups present in various PLA composite.

Composite	-C=O (% absorbance)	-C-O (% absorbance)	-CH3 (% absorbance)
PLA	0.0549	0.0023	0.0012
PLA/PVC	0.144	0.036	0.035
PLA/wood powder	0.117	0.0317	0.031
PLA/Fe_3O_4 powder	0.092	0.039	0.012
Hybrid-blended PLA	0.091	0.055	0.016

absorbance capacity of material matrix which meant with reinforcement present in base PLA matrix, the composite matrix hinders the path of light and absorbs the spectra passing through it thus showing poor application

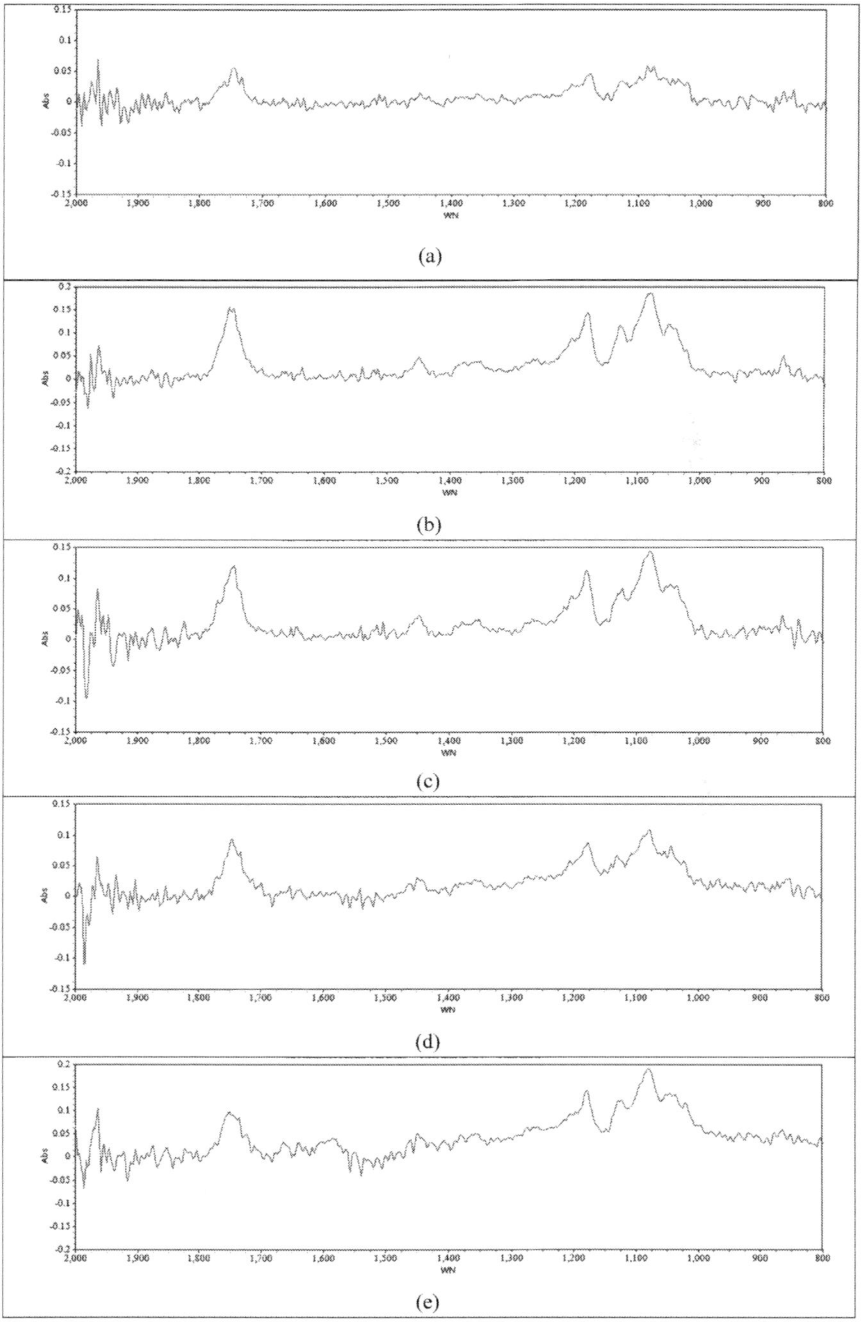

Figure 1.7 FTIR results for (A) pure PLA, (B) PLA/PVC, (C) PLA/Wood powder, (D) PLA/Fe_3O_4 powder, and (E) hybrid-blended PLA in the range of 800–2000 WN.

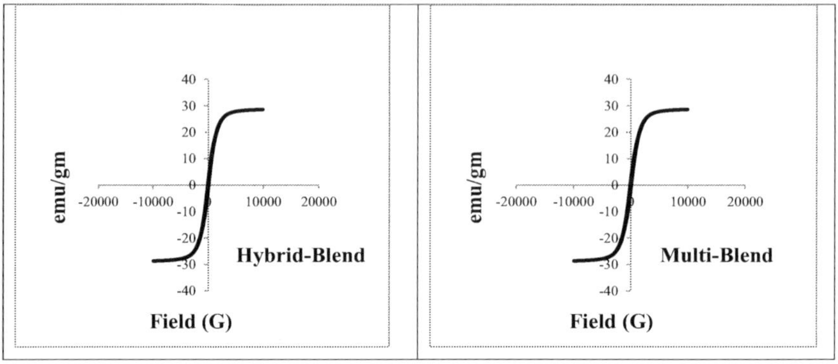

Figure 1.8 VSM hysteresis curve for hybrid and multimaterial matrix.

in case of remote sensing materials. It has been observed that with the increased absorbance the mechanical properties were seen to be reduced (Fig. 1.7). As it can be seen, the pure PLA has shown least absorbance and its mechanical properties were best among the different PLA composite [44]. Similarly, the PLA/Fe_3O_4 powder-based composite has shown second least absorbance due to which its mechanical properties were second best among the different PLA tested composites. It should be noted that the FTIR results are focused only for 800–2000 WN as above 2000 the absorbance was near to zero which meant that in range of 2000–4000WN the samples of PLA composite may have some remote sensing applications as they provide 99%–100% transmittance of signals.

1.7 Vibration sample magnetometery results

The samples of multimaterial and hybrid material composite of PLA were tested for magnetic characteristics using vibration sample magnetometery (VSM) machine in which an external magnetic field of 1Tesla was given to the samples and the extent of magnetization of sample was observed and from the observation it was made clear that the magnetization for both samples (hybrid and multimaterial) (Fig. 1.8) was purely temporary and on removal of external magnetic field the samples were again non-magnetic thus both samples had shown superparamagnetic characteristic as their hysteresis curve was very narrow.

1.8 Summary

The present investigation has dealt with the preparation of hybrid and multimaterial blend of PLA composite using wood powder, PVC, and

Fe_3O_4 powder as reinforcement so that results for 3D printed functional prototypes of different composition and combination may be compared for further 4D applications and following were the observations:

- The 3D printing of multimaterial and hybrid-blended functional prototype was feasible;
- The tensile and flexural properties of multimaterial-based 3D printed functional prototype were better than hybrid-blended matrix of PLA. It was observed that the multimaterial-based functional prototype have shown approximately 150% better results than hybrid blend of PLA;
- VSM-based analysis has shown that the magnetization of the multimaterial and hybrid material-based composite was similar which meant that both composites can show self-assembly characteristics on application of external magnetic field.
- FTIR-based analysis has highlighted that as the absorbance of the material increases with reinforcement level, the mechanical properties were reduced.

References

[1] Wohlers T, Gornet T. History of additive manufacturing. Wohlers Report 2014;24 (2014):118.

[2] Zhai Y, Lados DA, LaGoy JL. Additive manufacturing: making imagination the major limitation. JOM 2014;66(5):808–16.

[3] Tibbits S. 4D printing: multi-material shape change. Architectural Design 2014;84 (1):116–21.

[4] Ge Q, Dunn CK, Qi HJ, Dunn ML. Active origami by 4D printing. Smart Materials and Structures 2014;23(9):094007.

[5] Yu K, Dunn ML, Qi HJ. Digital manufacture of shape changing components. Extreme Mechanics Letters 2015;4:9–17.

[6] Srivastava V, Chester SA, Anand L. Thermally actuated shape-memory polymers: experiments, theory, and numerical simulations. Journal of the Mechanics and Physics of Solids 2010;58(8):1100–24.

[7] Hu J, Meng H, Li G, Ibekwe SI. A review of stimuli-responsive polymers for smart textile applications. Smart Materials and Structures 2012;21(5):053001.

[8] Gao B, Yang Q, Zhao X, Jin G, Ma Y, Xu F. 4D bioprinting for biomedical applications. Trends in Biotechnology 2016;34(9):746–56.

[9] Tibbits S, McKnelly C, Olguin C, Dikovsky D, Hirsch S. 4D printing and universal transformation. In: Proceedings of the 34th annual conference of the Association for Computer Aided Design in Architecture (ACADIA), Los Angeles, CA, 23–25 October, 2014, p. 539–548. ISBN 9781926724478.

[10] Pei E. 4D Printing: dawn of an emerging technology cycle. Assembly Automation 2014;34(4):310–14.

[11] Gladman AS, Matsumoto EA, Nuzzo RG, Mahadevan L, Lewis JA. Biomimetic 4D printing. Nature Materials 2016;15(4):413–18.

[12] Zhou Y, Huang WM, Kang SF, Wu XL, Lu HB, Fu J, et al. From 3D to 4D printing: approaches and typical applications. Journal of Mechanical Science and Technology 2015;29(10):4281–8.
[13] Raviv D, Zhao W, McKnelly C, Papadopoulou A, Kadambi A, Shi B, et al. Active printed materials for complex self-evolving deformations. Scientific Reports 2014;4:7422.
[14] Jamal M, Kadam SS, Xiao R, Jivan F, Onn TM, Fernandes R, et al. Bio-origami hydrogel scaffolds composed of photocrosslinked PEG bilayers. Advanced Healthcare Materials 2013;2(8):1142–50.
[15] Mao Y, Yu K, Isakov MS, Wu J, Dunn ML, Qi HJ. Sequential self-folding structures by 3D printed digital shape memory polymers. Scientific Reports 2015;5:13616.
[16] Wu J, Yuan C, Ding Z, Isakov M, Mao Y, Wang T, et al. Multi-shape active composites by 3D printing of digital shape memory polymers. Scientific Reports 2016;6:24224.
[17] Zhang Q, Zhang K, Hu G. Smart three-dimensional lightweight structure triggered from a thin composite sheet via 3D printing technique. Scientific Reports 2016;6:22431.
[18] Bakarich SE, Gorkin III R, Panhuis MI, Spinks GM. 4D printing with mechanically robust, thermally actuating hydrogels. Macromolecular Rapid Communications 2015;36(12):1211–17.
[19] Kuksenok O, Balazs AC. Stimuli-responsive behavior of composites integrating thermo-responsive gels with photo-responsive fibers. Materials Horizons 2016;3(1):53–62.
[20] Roy D, Cambre JN, Sumerlin BS. Future perspectives and recent advances in stimuli-responsive materials. Progress in Polymer Science 2010;35(1–2):278–301.
[21] Stuart MA, Huck WT, Genzer J, Müller M, Ober C, Stamm M, et al. Emerging applications of stimuli-responsive polymer materials. Nature Materials 2010;9(2):101–13.
[22] Sun L, Huang WM, Ding Z, Zhao Y, Wang CC, Purnawali H, et al. Stimulus-responsive shape memory materials: a review. Materials & Design 2012;33:577–640.
[23] Meng H, Li G. A review of stimuli-responsive shape memory polymer composites. Polymer 2013;54(9):2199–221.
[24] Whitesides GM, Grzybowski B. Self-assembly at all scales. Science (New York, N.Y.) 2002;295(5564):2418–21.
[25] Khademhosseini A, Langer R. A decade of progress in tissue engineering. Nature Protocols 2016;11(10):1775–81.
[26] Park JM, Kim DS. The influence of crystallinity on interfacial properties of carbon and SiC two-fiber/polyetheretherketone (PEEK) composites. Polymer Composites 2000;21(5):789–97.
[27] Kaboorani A, Riedl B. Nano-aluminum oxide as a reinforcing material for thermoplastic adhesives. Journal of Industrial and Engineering Chemistry 2012;18(3):1076–81.
[28] Aishima I, Takashi Y, Katayama Y, Arimoto K, Matsumoto K, inventors; Asahi Kasei Corp, assignee. Thermoplastic composite compositions. United States patent US 3,969,313, 1976.
[29] Goyal RK, Negi YS, Tiwari AN. High performance polymer composites on PEEK reinforced with aluminum oxide. Journal of Applied Polymer Science 2006;100(6):4623–31.
[30] Díez-Pascual AM, Naffakh M, Marco C, Gómez-Fatou MA, Ellis GJ. Multiscale fiber-reinforced thermoplastic composites incorporating carbon nanotubes: a review. Current Opinion in Solid State and Materials Science 2014;18(2):62–80.

[31] Pak SY, Kim HM, Kim SY, Youn JR. Synergistic improvement of thermal conductivity of thermoplastic composites with mixed boron nitride and multi-walled carbon nanotube fillers. Carbon 2012;50(13):4830–8.
[32] Pötschke P, Krause B, Buschhorn ST, Köpke U, Müller MT, Villmow T, et al. Improvement of carbon nanotube dispersion in thermoplastic composites using a three roll mill at elevated temperatures. Composites Science and Technology 2013;74:78–84.
[33] Cho EC, Huang JH, Li CP, Chang-Jian CW, Lee KC, Hsiao YS, et al. Graphene-based thermoplastic composites and their application for LED thermal management. Carbon 2016;102:66–73.
[34] Liu M, Papageorgiou DG, Li S, Lin K, Kinloch IA, Young RJ. Micromechanics of reinforcement of a graphene-based thermoplastic elastomer nanocomposite. Composites Part A: Applied Science and Manufacturing 2018;110:84–92.
[35] Yang X, Wang Z, Xu M, Zhao R, Liu X. Dramatic mechanical and thermal increments of thermoplastic composites by multi-scale synergetic reinforcement: carbon fiber and graphene nanoplatelet. Materials & Design 2013;44:74–80.
[36] Zarek M, Mansour N, Shapira S, Cohn D. 4D printing of shape memory-based personalized endoluminal medical devices. Macromolecular Rapid Communications 2017;38(2):1600628.
[37] Melocchi A, Inverardi N, Uboldi M, Baldi F, Maroni A, Pandini S, et al. Retentive device for intravesical drug delivery based on water-induced shape memory response of poly (vinyl alcohol): design concept and 4D printing feasibility. International Journal of Pharmaceutics 2019;559:299–311.
[38] Hoeher R, Raidt T, Krumm C, Meuris M, Katzenberg F, Tiller JC. Tunable multiple-shape memory polyethylene blends. Macromolecular Chemistry and Physics 2013;214:2725–32.
[39] Yu Y, Yang Y, Murakami M, Nomura M, Hamada H. Physical and mechanical properties of injection-molded wood powder thermoplastic composites. Advanced Composite Materials 2013;22(6):425–35.
[40] Nejhad MG, Parvizi-Majidi A. Impact behaviour and damage tolerance of woven carbon fibre-reinforced thermoplastic composites. Composites. 1990;21(2):155–68.
[41] Davies P, Cantwell WJ, Jar PY, Bourban PE, Zysman V, Kausch HH. Joining and repair of a carbon fibre-reinforced thermoplastic. Composites 1991;22(6):425–31.
[42] Kumar S, Singh R, Batish A, Singh TP. Additive manufacturing of smart materials exhibiting 4-D properties: a state of art review. Journal of Thermoplastic Composite Materials 2019. Available from: https://doi.org/10.1177/0892705719895052.
[43] Choksi N, Desai H. Synthesis of biodegradable polylactic acid polymer by using lactic acid monomer. International Journal of Applied Chemistry 2017;13(2):377–84.
[44] Kumar S, Singh R, Singh TP, Batish A. Investigations of polylactic acid reinforced composite feedstock filaments for multimaterial three-dimensional printing applications. Proceedings of the Institution of Mechanical Engineers, Part C: Journal of Mechanical Engineering Science 2019;233(17):5953–65.

CHAPTER 2

Graphene-reinforced acrylonitrile butadiene styrene composite as smart material for 4D applications

Vinay Kumar[1,2], Rupinder Singh[3] and I.P.S. Ahuja[2]
[1]Department of Production Engineering, Guru Nanak Dev Engineering College, Ludhiana, India
[2]Department of Mechanical Engineering, Punjabi University, Patiala, India
[3]Department of Mechanical Engineering, National Institute of Technical Teachers Training and Research, Chandigarh, India

2.1 Introduction

Polymers like acrylonitrile butadiene styrene (ABS), nylon, and polyethylene are nonconductive materials in nature due to which they are widely used as insulation on electrical equipment. But, it has been observed that the presence of reduced graphene (rG) in polymer matrix of ABS can make it electrically conductive for application of ABS-rG composite in manufacturing electrical instruments [1]. Electrochemical impedance spectroscopy of ABS-G platelet nanocomposite has also shown improvement in mechanical and electrical properties of ABS polymer-based composite matrix [2]. Some researchers have reported that multifunctional G-based nanocomposites of high-performance polymers matrix having ABS-molybdate-phosphorous-nitrogen-rG composition can be used efficiently for manufacturing high flame retardant and low smoke producing industrial products [3]. The modern world in which wearable devices and advance electronic components surrounds the environment, humans are more prone to harmful waves like electromagnetic interference; the highly compressed ABS-G foam-based sheets have been reported by some researchers that are capable of providing shield against such harmful radiations [4]. It has been reported that fused deposition modeling (FDM)-based 3D printed composite of ABS is a very useful and effective shield against electromagnetic radiations as ABS has good mechanical and thermal properties for 3D printing applications [5]. In order to improve

4D Printing
DOI: https://doi.org/10.1016/B978-0-12-823725-0.00004-7

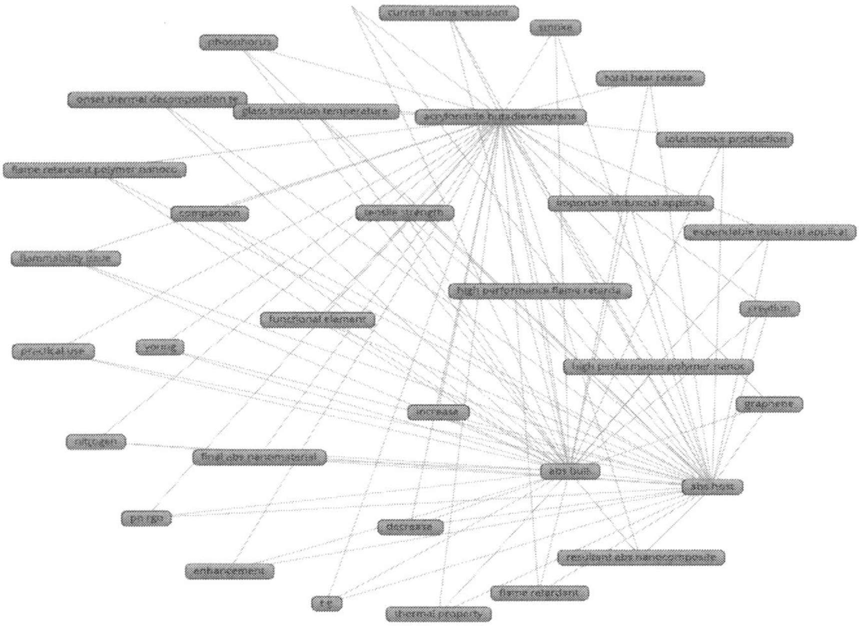

Figure 2.1 Web of terms investigated for research work performed on ABS. *ABS*, acrylonitrile butadiene styrene.

fatigue life of 3D printed prototypes of ABS matrix, it can be blended with polylactic acid (PLA) and by controlling the 3D printing direction fatigue failure can be control [6]. Fig. 2.1 shows the interaction of various research keywords like tensile strength, flame retardancy, composite, cryogenic treatment, etc. that have been explored on ABS and its composites by researchers around the globe (based upon web of science database of past 20 years).

The interaction of ABS, G, and other properties highlighted in Fig. 2.1 ensures that recently investigated polymer and hybrid composites with G present in them has been found good for antenna manufacturing, transistor, sensor, and many more industrial applications [7–10]. Table 2.1 shows the relevant score for various terms (used in research performed in past 20 years) obtained from web of science database for ABS and its composites.

Additively manufactured (AM) composites of ABS have been reported for some advance impedance characteristics and aeronautical applications against traditional commercial applications of ABS in manufacturing industries [11,12]. The large production activities of plastic products have

Table 2.1 List of experimentally investigated properties of acrylonitrile butadiene styrene (ABS) and its composites.

S. No.	Term	Occurrences	Relevance score
1.	ABS bulk	1	1.3965
2.	ABS host	1	1.3965
3.	Acceptable dynamic range	1	0.7696
4.	Acrylonitrile butadiene styrene	1	0.8363
5.	Acrylonitrile Butadiene styrene Copolymer	1	1.3965
6.	Basic framework	1	0.8363
7.	Calculated enthalpy	1	1.0098
8.	Catalytic performance enhancement	1	1.2509
9.	Comparative analysis	1	1.2509
10.	Composite	1	0.8363
11.	Contaminated E waste plastic	1	0.8363
12.	Critical temperature	1	1.0098
13.	Cryogenic temperature	1	1.0098
14.	Current flame retardant	1	1.3965
15.	E waste plastic	1	0.8363
16.	E waste plastics recycling	1	0.8363
17.	Electromagnetic field	1	1.219
18.	Electronic waste	1	0.8363
19.	Elevated temperature	1	0.8363
20.	Fiber optic sensing technique	1	1.219
21.	Field recycling	1	0.8363
22.	Final ABS nanomaterial	1	1.3965
23.	Flame retardant polymer nanocomposite	1	1.3965
24.	Flammability issue	1	1.3965
25.	Glass transition temperature	1	1.3965
26.	Graphene	1	1.3965
27.	High-performance flame retardant	1	1.3965
28.	High-performance polymer nanocomposite	1	1.3965
29.	Important industrial application	1	1.3965
30.	Matrix composition	1	0.7696
31.	Onset thermal decomposition temperature	1	1.3965
32.	Optical beam	1	1.2190
33.	Optical communication	1	1.2192
34.	Shape recovery	1	0.8363
35.	Recycling	1	0.8363
36.	Reference material	1	0.7696
37.	Temperature range	1	1.0098
38.	Tensile strength	1	1.3965
39.	Thermal degradation	1	0.8363
40.	Thermal processing procedure	1	0.7696

(Continued)

Table 2.1 (Continued)

S. No.	Term	Occurrences	Relevance score
41.	Thermal property	1	1.3965
42.	Thermal stress	1	0.8363
43.	Thermochemical process	1	0.8363
44.	Thermochemical treatment	1	0.8363
45.	Total heat release	1	1.3965
46.	Total smoke production	1	1.3965
47.	Transformation	1	1.2191

attracted the attention of researchers toward the least explored methods of recycling plastic waste so that methods of cleaner production activities can be explored. Various techniques [like cryo-milling, primary (1°), secondary (2°), and tertiary (3°) recycling] targeting waste management [13–18], hybrid/multimaterial development [19–22], and improving 3D printing capabilities of ABS and other thermoplastic waste have been reported so that innovative aspects of recycled polymeric materials can be implemented in modern-day manufacturing processes [23–25].

2.2 Research gap and problem formulation

The literature reported on experimentation conducted to explore the usefulness of ABS in past two decades shows that ABS is a low cost, convenient thermoplastic material to perform traditional and modern manufacturing practices (like 3D printing). ABS matrix-based composites with different reinforcements (like rG, G, carbon tubes, and carbon fibers) have increased the domain of 3D printing applications. But hitherto little has been reported on 4D capabilities of ABS composites for developing smart material with response toward external stimulus. Fig. 2.2 highlights the gap in the literature that G-reinforced ABS (GABS), ABS bulk, and ABS host have been investigated for 3D printing but GABS composites that possesses highly conductive and magnetic particles have not been investigated for 4D printing applications.

The present study reported the preparation of GABS composite GABS (by performing chemical-assisted mechanical blending (CAMB) of ABS plastic and nanoparticles of G) and investigating the 4D characteristics in various compositions/proportions of GABS. To test the 4D properties in GABS composites, shape memory effect was observed in wire samples

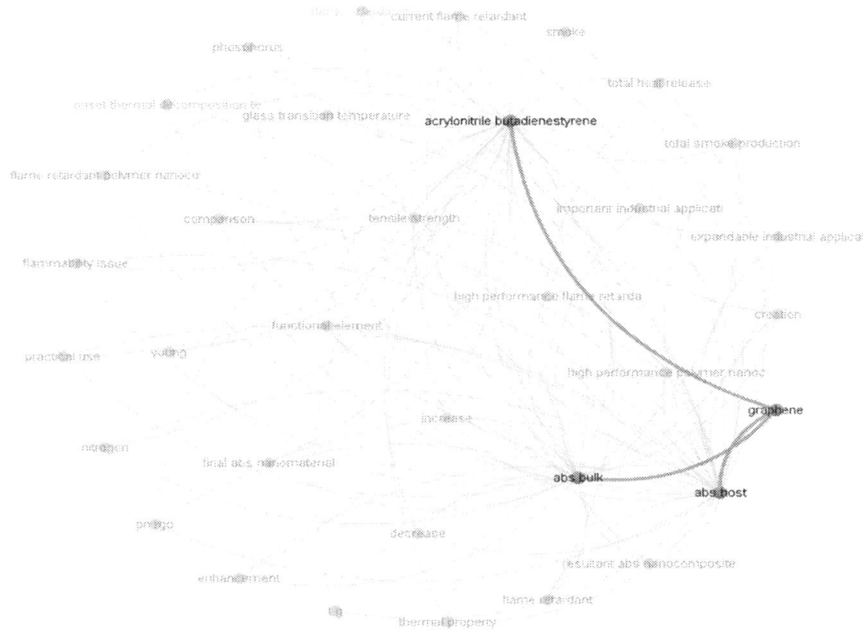

Figure 2.2 Research gap highlighting ABS-graphene composite for 4D applications. *ABS*, acrylonitrile butadiene styrene.

extruded using twin screw extruder (TSE) at optimized settings of the extruder setup. Further, vibration sample magnetometry (VSM) analysis and piezoelectric coefficient (D_{33}) calculations were performed to ascertain the magnetic and electric changes in GABS so that smart properties can be quantified in the 4D material. The morphological analysis was performed to investigate the effect of blending process and porosity in GABS wire samples on smart behavior of the proposed composite.

2.3 Experimentation

ABS granules, G powder, and solvent for ABS, that is, acetone was procured to perform CAMB of the components of the GABS composite. The weight proportion G varied from 0% to 20% by weight as the previous studies reported that best rheological, thermal, mechanical, and 3D printing properties can be achieved in GABS composites by consideration of the specified compositions of G in ABS polymer matrix [17]. Fig. 2.3 shows the work methodology followed for the 4D testing-based experimentation of GABS composite.

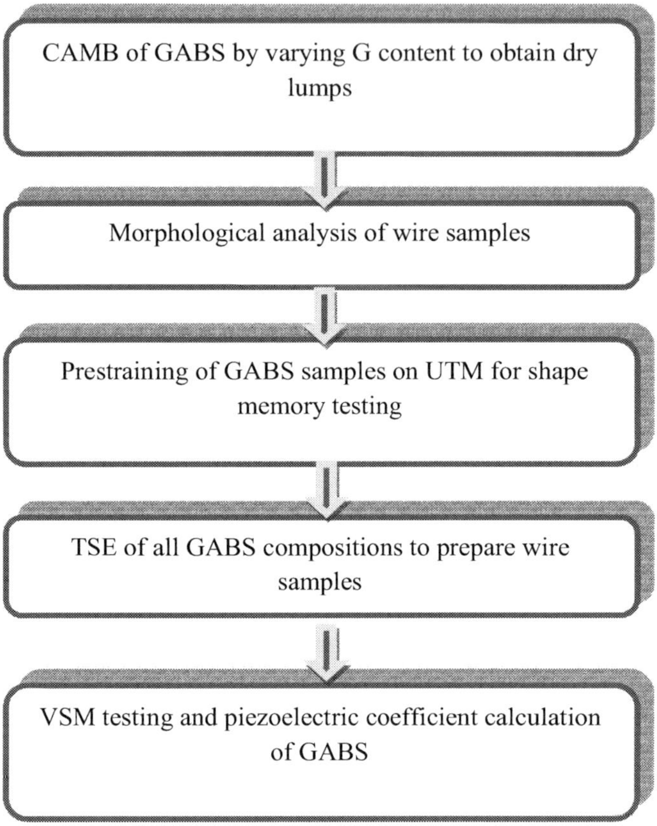

Figure 2.3 Work methodology for 4D characterization of GABS composite. *GABS*, G-reinforced ABS.

2.3.1 Chemical-assisted mechanical blending and TSE of G-reinforced ABS

Five compositions/proportions of GABS namely G0-ABS, G5-ABS, G10-ABS, G15-ABS, and G20-ABS were obtained in the form of dried lumps by evaporating acetone at 50°C present in the blend of GABS-acetone solution. Fig. 2.4A shows the lumps obtained for GABS composites. The dried lumps were fed to TSE shown in Fig. 2.4B. Wire specimens of each composition/proportion were extruded keeping screw temperature 210°C and 70 rpm screw speed. Fig. 2.4C shows five wire samples prepared for performing further experimentation.

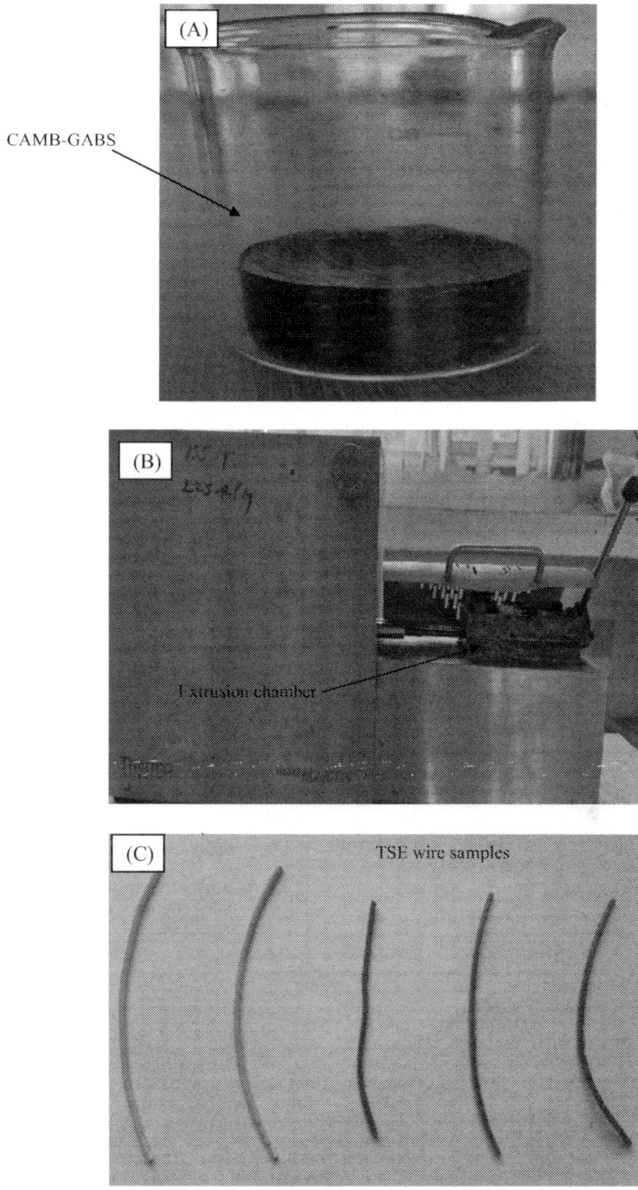

Figure 2.4 Dry lumps of GABS composite (A), TSE setup (B), and wire specimens of GABS for prestrain on UTM (C). *GABS*, G-reinforced ABS.

2.3.2 Prestraining G-reinforced ABS composite on universal testing machine

Approximately 100 mm long each GABS wire sample was allowed to remain in a vacuum oven for an hour whose temperature was maintained at 100°C (below glass transition temperature (T_g) of ABS) to release the post-TSE process stresses in the composite feedstock filament wires. Fig. 2.5A shows the vacuum oven used for heating GABS composite wires. After heat treatment in vacuum oven, wire specimens were strained for a controlled time (10 seconds) by keeping the speed of numeric control (NC) universal testing machine (UTM) (Fig. 2.5B) setup 20 mm/min. The change in length of each specimen was recorded, and they were

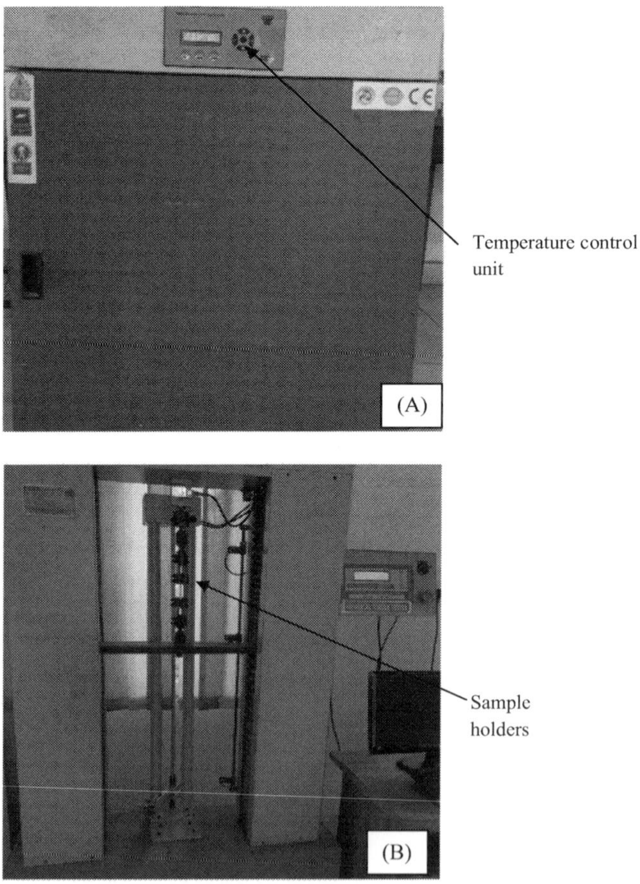

Figure 2.5 Vacuum oven (A) and NC UTM setup for straining GABS composite wire samples (B). *GABS*, G-reinforced ABS.

again held at 100°C temperature for 1 hour in order to record the final shape recovery in the composite. The objective was to study the effect of thermal environment around the GABS composite that acted as a stimulus to influence shape memory in the wire specimens.

2.3.3 Vibration sample magnetometry and piezoelectric analysis

Magnetic properties based on VSM analysis was performed for GABS compositions/proportions by application of 1Tesla external magnetic field. The magnetization induced in the GABS composite samples was recorded. 3D printed disk-shaped samples of GABS were poled using direct current (DC) poling unit and piezoelectric coefficient, D_{33} was recorded for each composition to ascertain volumetric changes in the 3D printed samples. Fig. 2.6 shows (A) VSM testing setup and (B) D_{33} meter.

2.4 Result and discussion

2.4.1 Shape memory effect in G-reinforced ABS composites

The results obtained for shape memory recovery in various GABS composite wire specimens is listed in Table 2.2. Below T_g, no specific phase transition occurred in the samples and significant change in specimen length was obtained.

Digital Vernier Calliper (Make: Mitutoyo, Japan) was used for measuring specimens length. It was observed that the increase in composition of G in GABS composites increased the shape memory effect of ABS. Highest recovery in length of G20-ABS wire sample was observed as maximum of 0.87 mm length of sample was recovered under stimulus effect of heat provided by vacuum oven at 100°C. On the other hand, negligible recovery was observed in G0-ABS in which no G was present. The outcomes of shape memory effect study show that GABS composites are highly thermally stable at room temperature but at elevated temperatures, it behaves like a heat responsive smart material with 4D characteristics.

2.4.2 Vibration sample magnetometry and piezoelectric analysis

The VSM and piezoelectric analysis results of each composition/proportion of GABS are listed in Table 2.3. The VSM analysis provided the maximum magnetization achieved (in terms of emu/g) in the composites

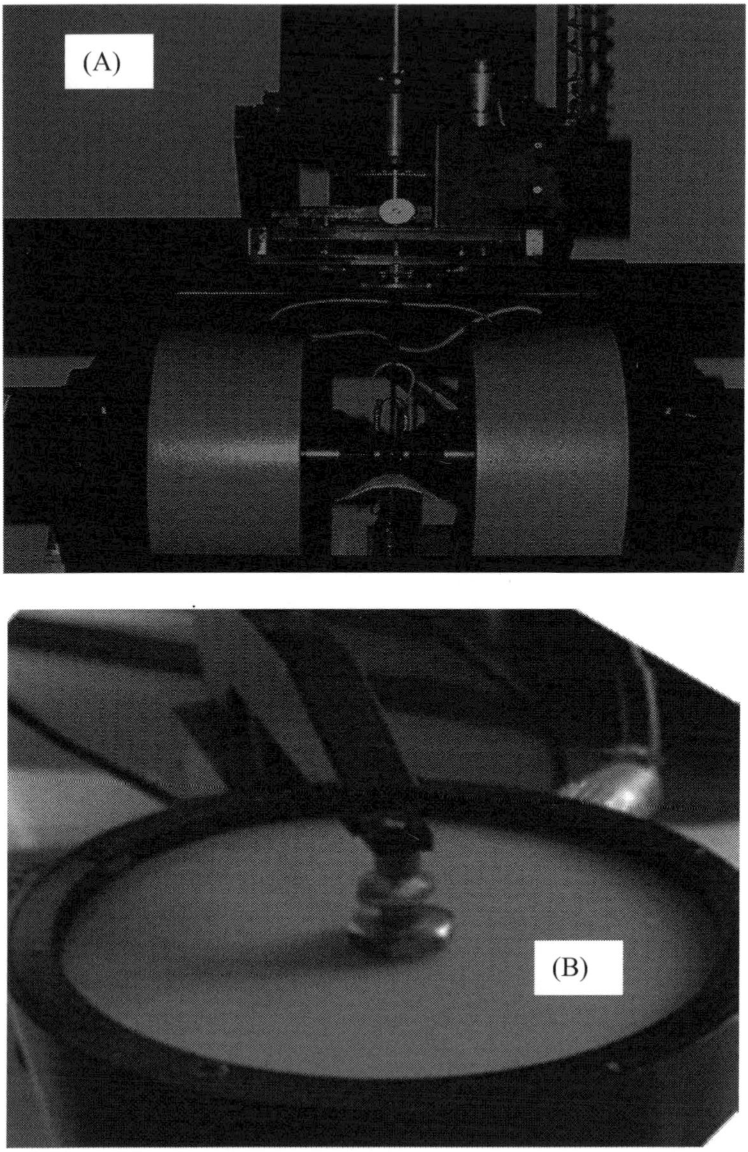

Figure 2.6 VSM testing machine (A) and piezoelectric coefficient (D_{33}) measuring setup (B). *VSM*, vibration sample magnetometry.

to outline 4D characteristics in GABS keeping external magnetic field as a function that provided magnetic properties to the material. Also, the piezoelectric coefficient observed in GABS shows the 4D properties in the material as a function of maximum possible volumetric change.

Table 2.2 Results obtained for shape memory recovery in GABS composite.

S. No.	Composition/ proportion	Initial specimen length (mm)	Change in length after strain (mm)	Specimen length after strain (mm)	Final specimen length on recovery (mm)	Length recovered (mm)
1.	G0-ABS	100.05	1.61	101.66	101.62	0.04
2.	G5-ABS	100.01	1.55	101.56	101.45	0.11
3.	G10-ABS	100.03	1.60	101.63	101.37	0.26
4.	G15-ABS	100.01	1.65	101.66	101.03	0.63
5.	G20-ABS	100.04	1.72	101.76	100.89	0.87

Table 2.3 VSM and piezoelectric coefficient results of GABS composites (as per Table 2.2).

S. No.	Composition/ proportion	Magnetization ($\times 10^{-5}$emu/g)	Possible capacity of DC poling (kV)	Piezoelectric coefficient D_{33} (pC/N)
1.	G0-ABS	0.0011	3	0.04
2.	G5-ABS	0.0216	3.15	1.19
3.	G10-ABS	0.0452	3.45	2.27
4.	G15-ABS	0.0571	3.50	3.61
5.	G20-ABS	0.0619	3.50	4.56

The results of the magnetic and piezoelectric study show that composite G20-ABS (having the highest blend of G) attained maximum magnetization of 0.0619×10^{-5}emu/g and piezoelectric coefficient ($D_{33} = 4.56$pC/N) with maximum DC poling capacity of 3.5 kV (without overloading) in comparison with rest of the samples of GABS. The composite shown acceptable smart properties in the presence of external applied stimulus (whether applied magnetic field and electric field) that changed the physical properties of the material.

2.4.3 Morphological analysis

The morphological properties of GABS wire samples were investigated (using Tool Maker's microscope at $\times 30$ magnification) along cross-sectional surface. Fig. 2.7 shows the porosity observed in wire samples along radial axis (using MIAS4.0 image analysis software). Based upon Fig. 2.7, Fig. 2.8 highlights the surface characteristics of the composite in

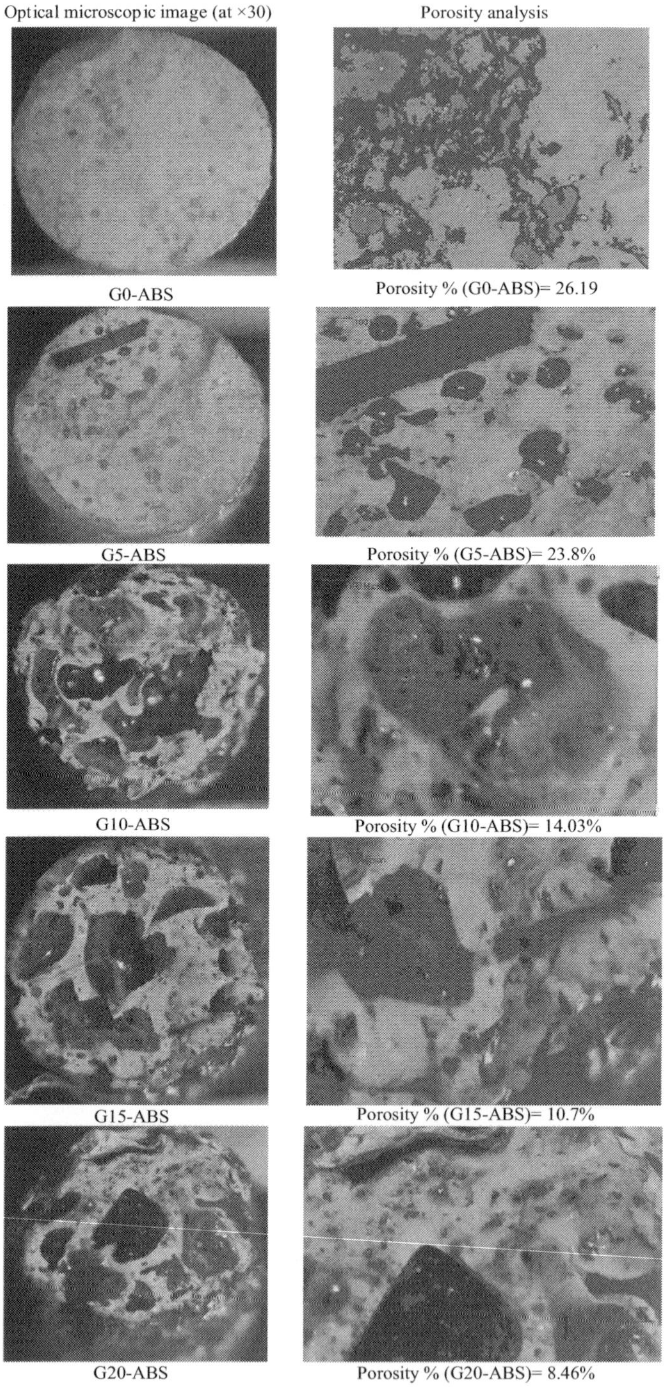

Figure 2.7 Optical microscopic image and porosity percentage of GABS composites (as per Table 2.2) along the radial axis. *GABS*, G-reinforced ABS.

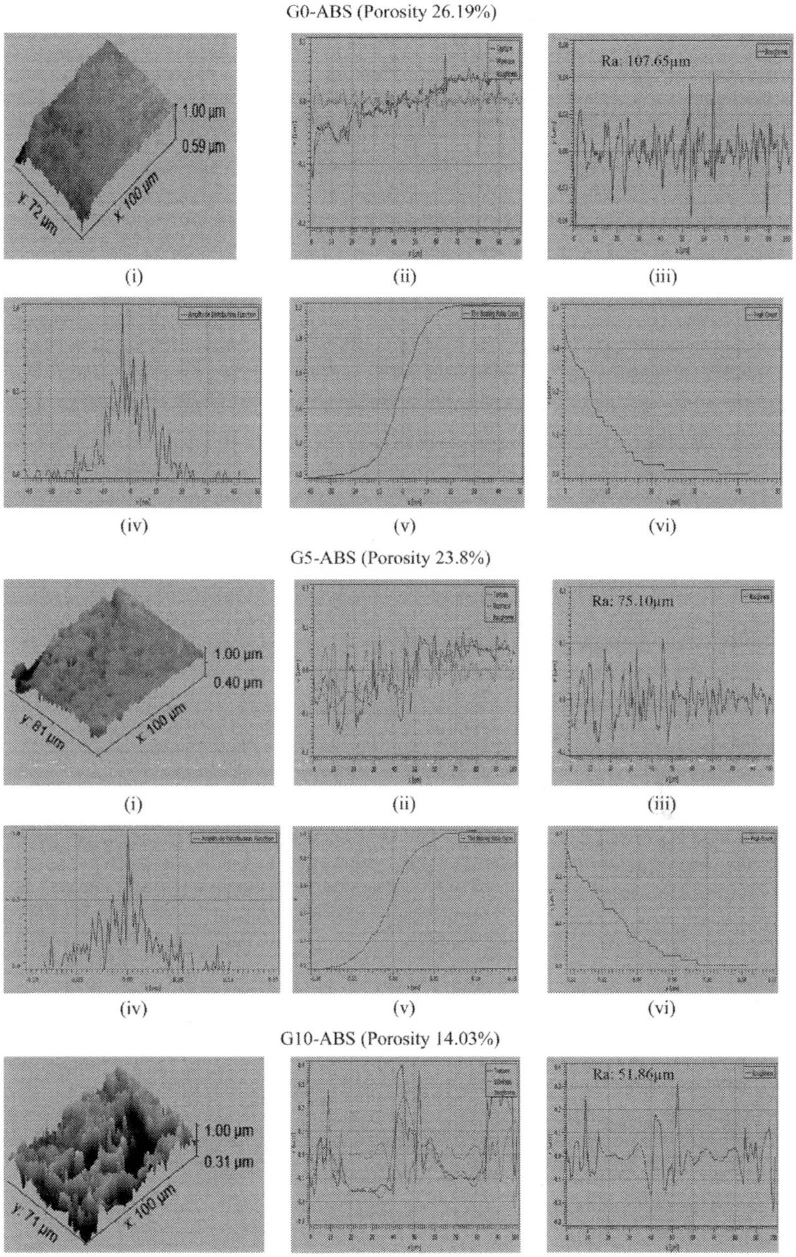

Figure 2.8 Surface characteristics along the radial surface of ABS and GABS composite samples: (i) 3D rendered image, (ii) surface texture, waviness profile, (iii) surface roughness profile, (iv) amplitude distribution function, (v) bearing ratio curve, and (vi) peak count. *ABS*, acrylonitrile butadiene styrene.

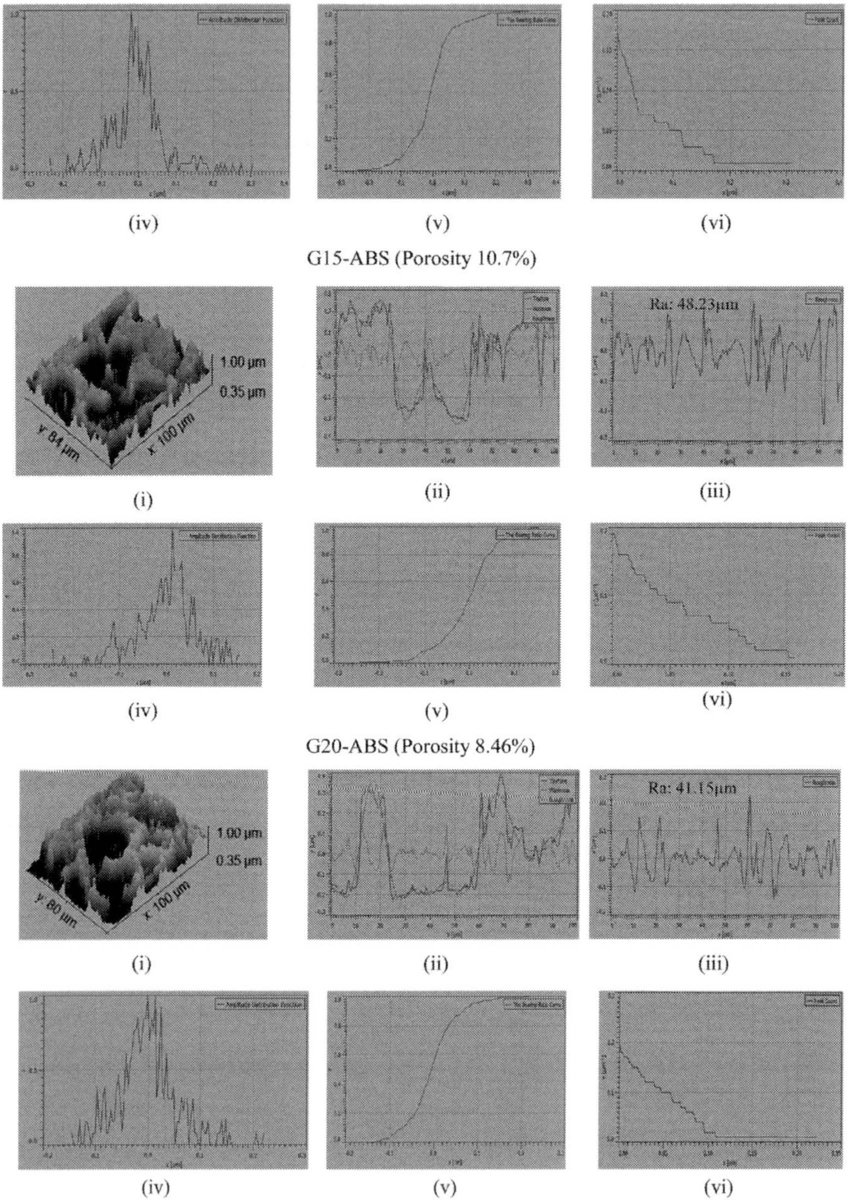

Figure 2.8 (*Continued*).

terms of surface texture, roughness, amplitude distribution function, peak count, bearing ratio curve.

From Fig. 2.7, Table 2.2, and Table 2.3, it has been ascertained that the controlled porosity for sample at S. No.5 (porosity 8.46%) in G20-ABS wire cross section played a significant role in imparting 4D properties. The uniform blending of material may have led to better affinity of composite particles and less porosity in feedstock filament wire. Hence, better shape memory effect was observed in G20-ABS. Also, quality of 3D printed sample was improved due to better morphological characteristics as a result of which chances of charge leakage got reduced and good VSM and piezoelectric properties were achieved. The optical photomicrographs (Fig. 2.7) after processing with image processing software clearly justify this affect in 3D rendered image, surface texture, waviness, surface roughness profile, amplitude distribution function, peak count, and bearing ratio curve analysis (Fig. 2.8).

Summary

The outcomes of present work may be concluded as under:

- The CAMB has been observed as effective process to prepare in-house smart composite of ABS for lab scale experimentation of 3D/4D applications of ABS-based polymer composite matrix. The CAMB-GABS composites are easy to process on TSE for fabrication of in-house developed 3D printing feedstock filaments.
- Good thermal responsive properties in GABS composites were responsible to impart significant shape memory effect in them. G20-ABS can be considered as the best composition since it has shown good shape memory effect with recovery of 0.87 mm length under the stimulus effect of heat.
- The VSM and piezoelectric analysis of GABS outlined acceptable magnetic and electro-active properties in the compositions/proportions of GABS in regard to 4D properties. G20-ABS shows magnetization of 0.0619×10^{-5}emu/g and piezoelectric coefficient $D_{33} = 4.56$pC/N. These properties are responsible for self-actuation in the material under the application of applied magnetic or electric field.
- The results of morphological analysis also supported the research outcomes as very less porosity was observed in G20-ABS sample along the radial axis. The porosity of 8.46% was observed for this composition/proportion, which may have contributed for better 4D properties.

Future scope

The GABS composite can be used as a smart material in 4D applications where thermal responsive, small size thermoplastic-based components are installed for controlled actuation and retraction. The VSM and piezoelectric-based 4D capabilities of GABS may be explored to develop smart 3D printed prototypes that possess self-actuating and self-healing properties for robotics sensors, actuators, and new era smart antennas for various engineering applications.

Acknowledgment

The authors received financial support under Government of India initiative Science and Heritage Research Initiative (SHRI) project to develop smart material for repairing damaged heritage structures under File No. DST/TDT/SHRI-35/2018.

References

[1] Gao A, Zhao F, Wang F, Zhang G, Zhao S, Cui J, et al. Highly conductive and light-weight acrylonitrile-butadiene-styrene copolymer/reduced graphene nanocomposites with segregated conductive structure. Composites Part A: Applied Science and Manufacturing 2019;122:1–7.

[2] Alauddin SM, Ismail I, Zaili FS, Ilias NF, Aripin NFK. Electrical and Mechanical properties of Acrylonitrile butadiene styrene/graphene platelet nanocomposite. Materials Today: Proceedings 2018;5:S125–9.

[3] Huang G, Chen W, Wu T, Guo H, Fu C, Xue Y, et al. Multifunctional graphene-based nano-additives toward high-performance polymer nano-composites with enhanced mechanical, thermal, flame retardancy and smoke suppressive properties. Chemical Engineering Journal 2021;410:127590.

[4] Wang L, Wu Y, Wang Y, Li H, Jiang N, Niu K. Laterally compressed graphene foam/acrylonitrile butadiene styrene composites for electromagnetic interference shielding. Composites Part A: Applied Science and Manufacturing 2020;133:105887.

[5] Schmitz DP, Dul S, Ramoa SDAS, Soares BG, Barra GMO, Pegoretti A. Effect of printing parameters on the electromagnetic shielding efficiency of ABS/carbonaceous-filler composites manufactured via filament fused fabrication. Journal of Manufacturing Processes 2021;65:12–19.

[6] Azadi M, Dadashi A, Dezianian S, Kianifar M, Torkaman S, Chiyani M. High-cycle bending fatigue properties of additive-manufactured ABS and PLA polymers fabricated by fused deposition modeling 3D-printing. Forces in Mechanics 2021;100016.

[7] Clower W, Hartmann MJ, Joffrion JB, Wilson CG. Additive manufactured graphene composite Sierpinski gasket tetrahedral antenna for wideband multi-frequency applications. Additive Manufacturing 2020;32:101024.

[8] Jyoti J, Dhakate SR, Singh BP. Phase transition and anomalous rheological properties of graphene oxide-carbon nanotube acrylonitrile butadiene styrene hybrid composites. Composites Part B: Engineering 2018;154:337–50.

[9] Tao T, Zhou Y, Ma M, He H, Gao N, Cai Z, et al. Novel graphene electrochemical transistor with ZrO2/rGO nano-composites functionalized gate electrode for ultrasensitive recognition of methyl parathion. Sensors and Actuators B: Chemical 2021;328:128936.

[10] Nabi G, Malik N, Tahir MB, Raza W, Rizwan M, Maraj M, et al. Synthesis of graphitic carbon nitride and industrial applications as tensile strength reinforcement

agent in red Acrylonitrile-Butadiene-Styrene (ABS). Physica B: Condensed Matter 2021;602:412556.

[11] Jyoti J, Kumar A, Dhakate SR, Singh BP. Dielectric and impedance properties of three dimension graphene oxide-carbon nanotube acrylonitrile butadiene styrene hybrid composites. Polymer Testing 2018;68:456–66.

[12] Oztan C, Ginzburg E, Akin M, Zhou Y, Leblanc RM, Coverstone V. 3D printed ABS/paraffin hybrid rocket fuels with carbon dots for superior combustion performance. Combustion and Flame 2021;225:428–34.

[13] Cress AK, Huynh J, Anderson EH, O'neill R, Schneider Y, Keleş Ö. Effect of recycling on the mechanical behavior and structure of additively manufactured acrylonitrile butadiene styrene (ABS). Journal of Cleaner Production 2021;279:123689.

[14] Kumar V, Singh R, Ahuja IPS. On mechanical and thermal properties of cryo-milled primary recycled ABS. Sādhanā 2020;45(1):1–13.

[15] Kumar V, Singh R, Ahuja IPS. On cryogenic milling of primary recycled ABS: rheological, morphological, and surface properties. Journal of Thermoplastic Composite Materials 2020. p. 0892705720932621.

[16] Kumar V, Singh R, Ahuja IPS. Effect of extrusion parameters on primary recycled ABS: mechanical, rheological, morphological and thermal properties. Materials Research Express 2020;7(1):015208.

[17] Kumar V, Singh R, Ahuja IPS. Secondary recycled acrylonitrile–butadiene–styrene and graphene composite for 3D/4D applications: rheological, thermal, magnetometric, and mechanical analyses. Journal of Thermoplastic Composite Materials 2020. p. 0892705720925114.

[18] Singh PK, Suman SK, Kumar M. Influence of Recycled Acrylonitrile Butadiene Styrene (ABS) on the Physical. Rheological and Mechanical Properties of Bitumen Binder. Transportation Research Procedia 2020;48:3668–77.

[19] Singh R, Kumar R, Singh P. Prospect of 3D printing for recycling of plastic product to minimize environmental pollution. Reference Module in Materials Science and Materials Engineering. Elsevier; 2018.

[20] Singh R, Kumar R. Energy storage device from polymeric waste based nano-composite by 3D printing. Reference Module in Materials Science and Materials Engineering. Elsevier; 2020.

[21] Kumar S, Singh R, Singh TP, Batish A. Thermosetting polymers for 4D printing. Reference Module in Materials Science and Materials Engineering. Elsevier; 2020.

[22] Kumar S, Singh R, Singh TP, Batish A. Dual/Multi printing of thermosetting polymers. Reference Module in Materials Science and Materials Engineering. Elsevier; 2020.

[23] Kumar S, Singh R, Singh TP, Batish A. Multi material printing of recycled thermoplastics and thermosetting polymers. Reference Module in Materials Science and Materials Engineering. Elsevier; 2020.

[24] Singh J, Chawla K, Singh R. Mechanical and rheological investigations of bakelite reinforced ABS. Reference Module in Materials Science and Materials Engineering. Elsevier; 2020.

[25] Rabbi MF, Chalivendra V. Strain and damage sensing in additively manufactured CB/ABS polymer composites. Polymer Testing 2020;90:106688.

CHAPTER 3

Two-way programming of secondary recycled poly(lactic) acid composite matrix using magnetic field as stimulus

Sudhir Kumar[1], Rupinder Singh[2], T.P. Singh[3] and Ajay Batish[3]
[1]Department of Mechanical Engineering, CT University, Ludhiana, India
[2]Department of Mechanical Engineering, National Institute of Technical Teachers Training and Research, Chandigarh, India
[3]Department of Mechanical Engineering, Thapar Institute of Engineering and Technology, Patiala, India

3.1 Introduction

The smart material matrices in three-dimensional (3D) printing are one of the most important fields of research these days. The introduction of smart material matrices in 3D printing has led the development of 3D printing to the four-dimensional (4D) printing [1]. The fourth dimension of 4D printing is "time" [2]. The smart material matrix is also known as shape memory polymers. These material matrices are termed as smart as they are able to memorize their original shape and can change the shape on application of some external stimulus [3]. In recent years, different studies on multimaterial printings have been reported the 4D effect of material matrix based on design of specimen for example, miura-origami structure and other meta material structures [4–6]. The meta material structures are designed in such a way that when external stimulus is applied, their geometry leads to change in their structure [7,8]. The important geometrical details which led to change in structure are edge length, join angle, curve shapes, etc.

4D printing has as an idea evolved by Massachusetts Institute of Technology (MIT) in 2013 where the MIT lab demonstrated the possibility of 4D behavior. The basic idea illustrated that the material of 3D printing can mimic the different natural phenomenon already existing in the environment [9]. There exist different ways of material programming printing such as one-way programmed, two-way programmed, and multi-way programming of material matrix for 4D applications. Fig. 3.1 shows

4D Printing
DOI: https://doi.org/10.1016/B978-0-12-823725-0.00006-0

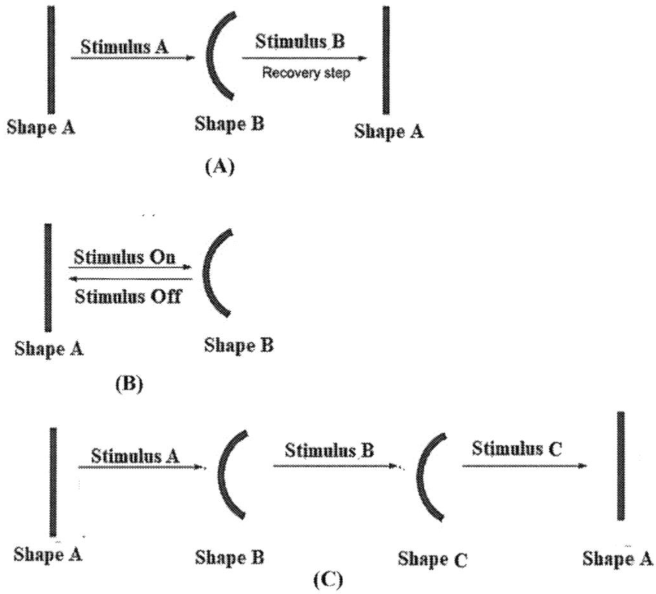

Figure 3.1 Different ways of programming (A) single-way, (B) two-way, and (C) multiway.

the one-way, two-way, and multiway programming techniques. One-way programming means the material matrix or design is coded for single stage of shape change and it required different stimulus at different time to acquire the required change in shape [10]. The two-way programming means the entity can get reverse back to original form with the same stimulus by making the stimulus on and off whenever required [11]. Multiway programming deals with the shape recovery and shape change for different stages (greater than two) [12,13].

Various stimuli have been explored by researchers in recent past (such as pH value, temperature, sunlight, electrical field, and magnetic field) [14]. The existence of different crystal structures such as martensite, twinned martensite, and austenite, etc. enhances the possibility of different structural changes in shape memory alloys [12]. The hysteresis for the smart material plays an important role in its application as low hysteresis is necessary for fast actuation [15]. In case of magnetic material, the twin boundary motion is the basis of shape memory effect. The magnetically induced reorientation about the twin boundary causes the structural changes in the material matrix [16].

With the increase in the temperature of surroundings, the 4D behavior of material reduces as the increase in temperature leads to loss in magnetization [17]. The use of 4D printing of structural materials may lead to large-scale benefits to the different industry such as medical industry may be benefited by different 4D printed materials. For example, use of NiTi alloys for self-correction of teeth and self-expanding stunts in biomedical applications [18,19].

3.2 Two-way programming of secondary recycled poly (lactic)acid composite: a case study

The literature review revealed about the changing phase of 3D printing into 4D printing. The 4D printing required material programming for particular stage and an external stimulus for actuation of the material matrix. In present research work, a case study has been performed using recycled poly(lactic)acid (PLA) composite matrix in which Fe_3O_4 (magnetite) powder has been reinforced so that effect of magnetic field can be observed and changes in the behavior may be observed using vibration sample magnetometery (VSM). In the end, the magnetization capability of the developed feedstock was tested for its repeatability and reliability by calculating its statistical control performance over the different recycling stages of polymeric composite.

3.3 Materials and method

For present research work, PLA polymer and Fe_3O_4 powder have been purchased from local market (Shiva Chemicals Limited, Ludhiana, India). Mechanical blending method was used to prepare the composite of PLA. The proportion of Fe_3O_4 powder was 20% in PLA matrix, and it was selected on the basis of melt flow index of material matrix. The developed composite matrix was then subjected to twin screw extrusion (TSE) process for feedstock filament development. The developed feedstock filament was then again reextruded using TSE machine for 3 number of times, and samples of feedstock filament were collected for different stage of recycling. The recycling was performed for three stages so that effect of the recycling on different properties may be observed. The developed feedstock was tested for mechanical and magnetic properties using universal tensile testing (UTM) machine and VSM machine respectively. Fig. 3.2 shows the methodology used for the present case study.

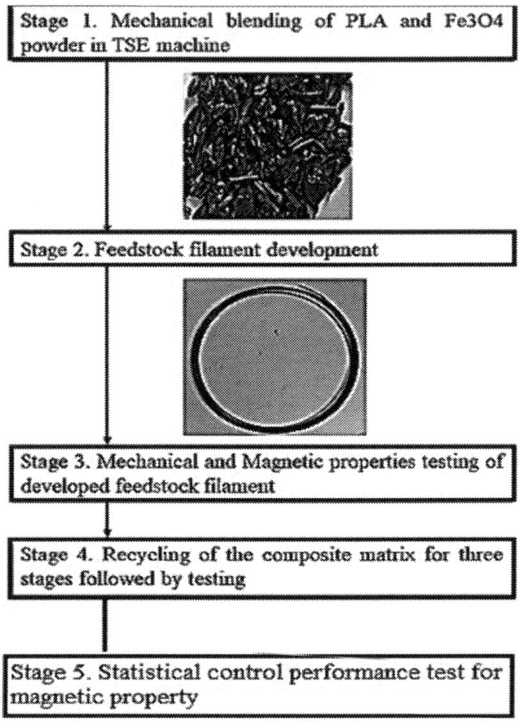

Figure 3.2 Methodology followed in present case study.

3.4 Result and discussion

3.4.1 Mechanical testing results

The developed feedstock filaments for different stages of recycling were tested for mechanical properties using UTM machine. Fig. 3.3 shows the mechanical performance of the PLA/Fe_3O_4 composite material matrix with different stages of recycling. It was observed that with the increase in the recycling stages for PLA/Fe_3O_4 composite, the mechanical performance was decreased but the change in the mechanical performance for different properties was not significant. The loss in the peak strength of PLA/Fe_3O_4 matrix over the three recycling stages was nominal as it can be seen that the loss in peak strength for first stage it was 1%; for the second stage it was 14%; and for the third stage it was 22%. Thus the reinforcement of Fe_3O_4 powder in PLA matrix improves the recycling life of PLA as well as introduces magnetic characteristics in the material matrix which was necessary as for the 4D nature of the smart material matrix magnetic characteristics were required.

3.4.2 Vibration sample magnetometery analysis

The feedstock filament was tested for its magnetic characteristics using VSM machine using room temperature conditions. The test result suggested that the sample held superparamagnetic characteristics as the hysteresis loop was very thin. Thin hysteresis loop suggests that the material may be used with fast actuation and deactuation process as the material matrix held low retentivity and coercivity. The magnetization value for the feedstock filament was observed to be 28 emu/g for the selected composite ratio. Table 3.1 shows the magnetic properties observed for the prepared feedstock filament from which it was clear that there was very minute change in the magnetic properties which was not significant.

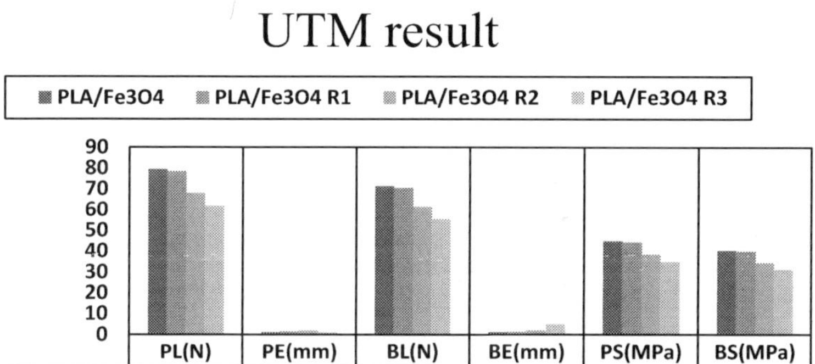

	PL(N)	PE(mm)	BL(N)	BE(mm)	PS(MPa)	BS(MPa)
PLA/Fe3O4	79.4	1.14	71.46	1.33	44.95	40.46
PLA/Fe3O4 R1	78.4	1.52	70.56	1.52	44.39	39.95
PLA/Fe3O4 R2	68.1	1.9	61.29	2.09	38.56	34.7
PLA/Fe3O4 R3	61.7	1.14	55.53	5.13	34.93	31.44

Figure 3.3 Mechanical results for PLA composite for different recycling stages. Note: *PL*, peak load; *PE*, peak elongation; *BL*, break load; *BE*, break elongation; *PS*, peak strength; *BS*, break strength.

Table 3.1 Magnetic properties observed for the recycled material matrix for the three stages.

Composite/recycling stage	Magnetization (emu)	Retentivity
PLA/Fe_3O_4	28.553 ± 1.2	72.354 ± 4.32
PLA/Fe_3O_4 R1	27.869 ± 1.4	71.263 ± 3.56
PLA/Fe_3O_4 R2	27.965 ± 1.3	72.243 ± 3.21
PLA/Fe_3O_4 R3	28.114 ± 1.2	72.361 ± 4.14

Fig. 3.4 shows the hysteresis loop for the PLA/Fe_3O_4 (80/20 wt.%). The hysteresis loop for the three stages of recycling has shown that there was marginal change in the magnetic characteristics of PLA/Fe_3O_4 composite as the material matrix was processed with low-temperature condition of 210 °C in TSE machine setup which was not enough to change the magnetic characteristics of the material. The VSM analysis has given clear indication that the samples were of superparamagnetic structure as it may be observed from hysteresis loop area. Therefore, the material can be magnetized and demagnetized at very fast rate. Thus two-way programming of the feedstock filament material prepared by using TSE process was possible. Three samples were made for each type of composition so that statistical control of process may be checked.

3.4.3 Statistical control of magnetic properties of poly(lactic) acid composites

Table 3.1 shows the advanced excel software in which statistical control was calculated for the different PLA composite. Table 3.2 shows the upper and lower standard limits to perform statistical control test of the obtained data. The upper statistical and lower statistical limit for the

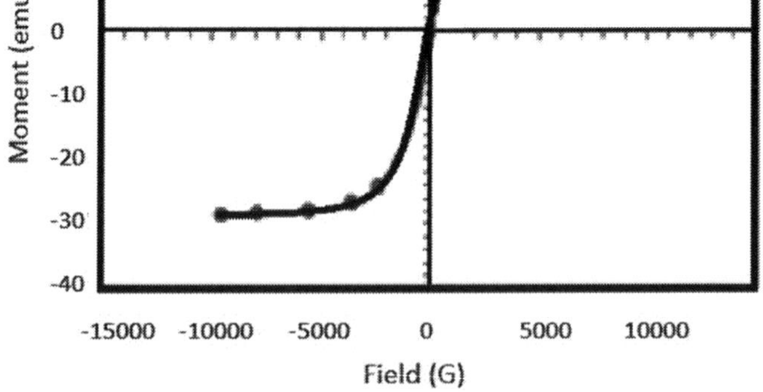

Figure 3.4 Hysteresis loop for the PLA/Fe_3O_4 (80/20 wt%). Note: The hysteresis loop for the other recycled material matrix is not shown as the material matrix held the same result and adding the data for the remaining material matrixes would result in overlapping of the curve.

Table 3.2 Upper and lower statistical control limit for PLA/Fe_3O_4 composite for three stages of recycling.

Upper statistical control limit	29
Lower statistical control limit	27

statistical control were selected on the basis of maximum and minimum result obtained for the material composition without recycling such as in case of PLA/Fe_3O_4 composite for three stages of recycling the magnetic properties of the feedstock filament prepared at stage 1 was used as upper and lower control limits. From the results of statistical control test, it may be observed that the process was in statistical control as its C_p and C_{pk} values were above 1 and standard deviation was optimum for the PLA/Fe_3O_4 composite. The result also suggests that the different feedstock of different recycling stages can easily be magnetized and demagnetized as the value was repeating and was in control limit for each case. Thus it may be established that the two-way programming of PLA/Fe_3O_4 composite is feasible and repeatable for three stages of recycling. Fig. 3.5A and B shows the normality probability test graph and bell shape curve, respectively, for magnetization of PLA/Fe_3O_4 composite from which it may be observed that the magnetization of the composite was in statistical control limit. Table 3.3 shows overall and potential capability of the process of feedstock filament production focusing on magnetic properties of material composite which includes C_p and C_{pk} values for the process.

3.4.4 Porosity analysis

The different feedstock filaments of PLA-based composites were analyzed for surface characteristics under Tool maker's microscope using $\times 100$ scale. It was observed that with the increasing recycling stages, the material matrix became porous and there was increment in the voids/porosity holes of the material matrix. Fig. 3.6 shows the photomicrographs of the developed feedstock filament of PLA/Fe_3O_4 composite. From porosity results, it was clear that the PLA/Fe_3O_4 composite held lowest porosity (5%) percentage and sample 4 (PLA/Fe_3O_4 R3) held largest porosity (24%) level among all.

3.4.5 3D surface rendering and surface roughness analysis

The photomicrographs from Tool maker's microscope were further analyzed using open-source 3D rendering software. From surface rendering,

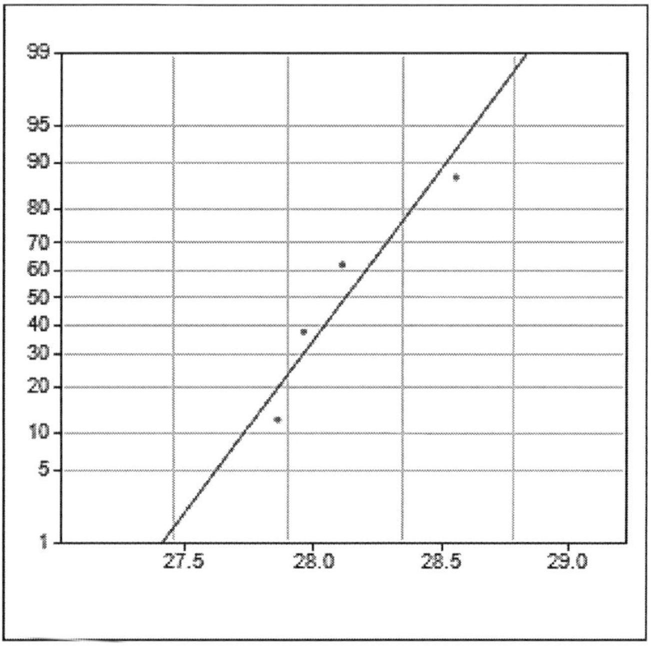

(A)

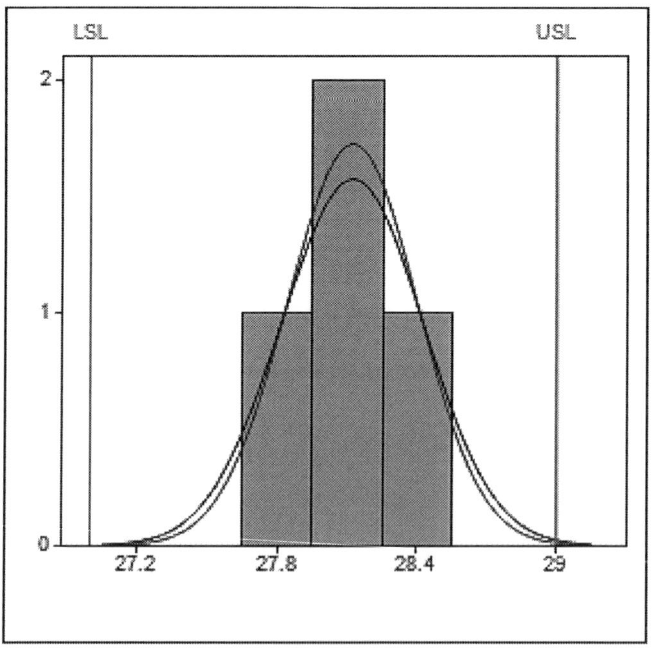

(B)

Figure 3.5 (A) Normality probability test graph and (B) bell shape curve for magnetization.

Table 3.3 Overall and potential capability of the process of feedstock filament production focusing on magnetic properties of material composite.

Potential capability		Overall capability	
Std. deviation	0.277778	Std. deviation	0.304521
C_p	1.2	P_p	1.09
C_{pk}	1.06	P_{pk}	0.96

Note: The difference between C_p and P_p is how the std. deviation is determined. For P_p, sampling was used to calculate an estimated standard deviation of the sample. In C_p, stable process was assumed to calculate a true standard deviation.

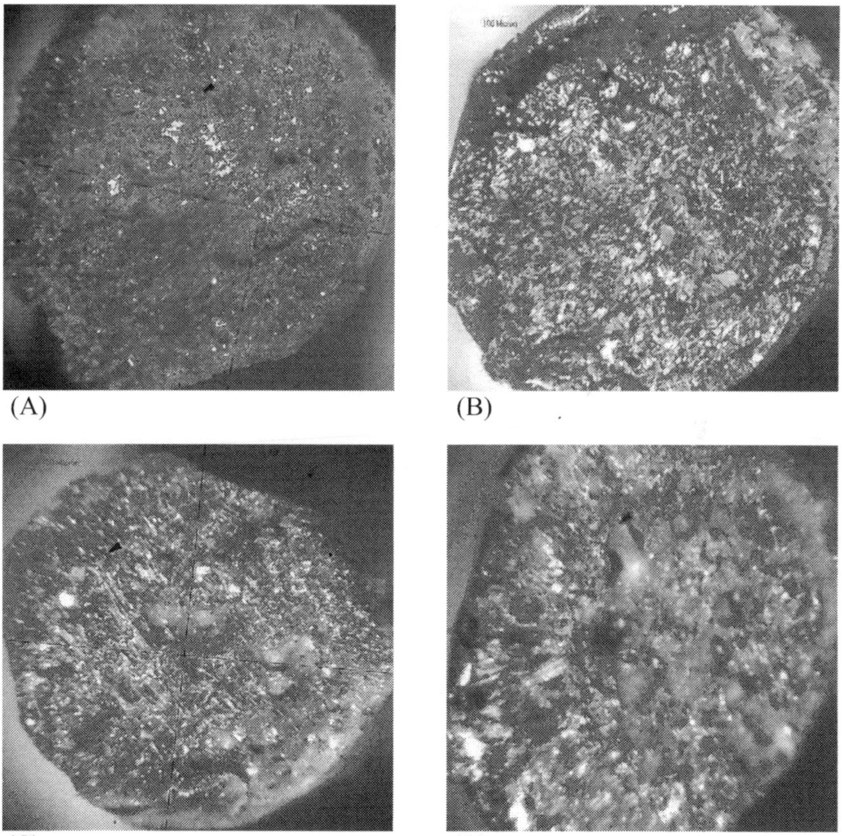

(A) (B) (C) (D)

Figure 3.6 Photomicrographs of feedstock filaments at × 100: (A) percentage porosity PLA/Fe_3O_4: 5%; (B) percentage porosity PLA/Fe_3O_4 R1: 9%; (C) percentage porosity PLA/Fe_3O_4 R2: 15%; (D) percentage porosity PLA/Fe_3O_4 R3: 24%.

3D rendered image and surface roughness profile for different feedstock filament was obtained (Fig. 3.7). From surface roughness profile, it was observed that the sample 1 with least porosity held least surface roughness

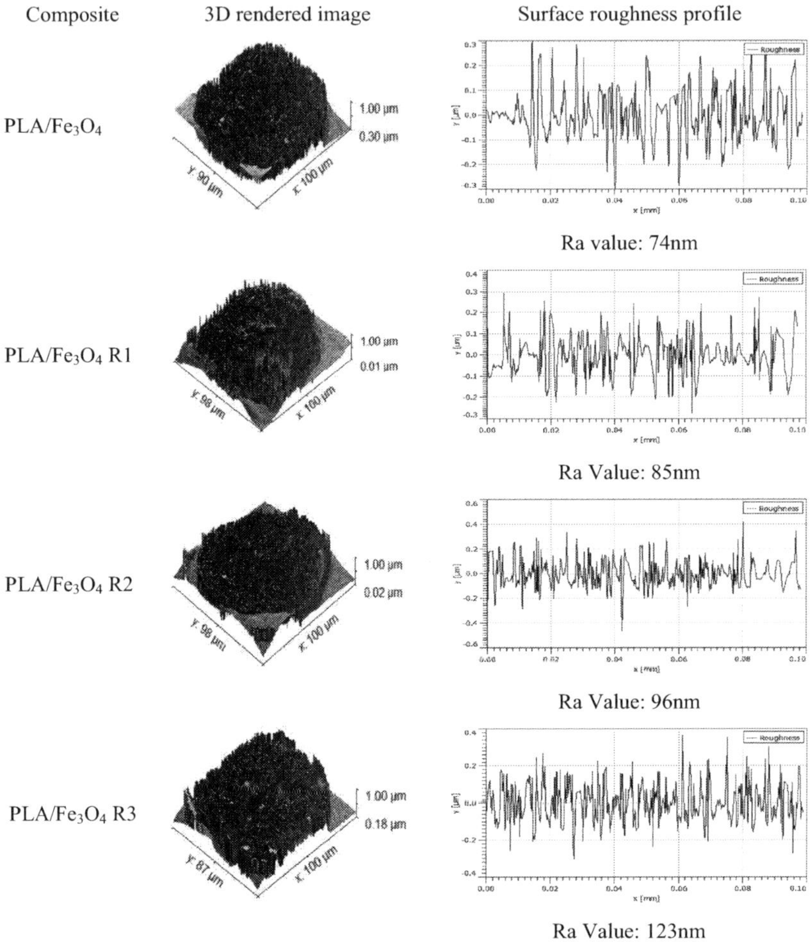

Figure 3.7 3D rendered image and surface roughness profile for different recycled PLA composites.

(Ra value 74 nm), whereas the sample 4 with large porosity holes has shown maximum surface roughness (Ra value 123 nm).

3.5 Studies reported at international level

In order to ascertain the studies performed at international level in past two decades, web of science database has been explored on two-way programming of secondary recycled PLA composite matrix using magnetic field as stimulus for 4D applications. The results of study outlined that limited studies are available for programming of recycled PLA composite matrix. However, significant studies (572 articles) have been reported on recycled PLA (from 1999–2021). In web of science, core collection database by fixing minimum number of occurrences of term "5," out of 1814 terms, 56 terms met the threshold. Out of these 56 terms, relevance score was calculated (Annex. 1-(Relevance score with occurrences for keyword recycled PLA) Table 3.A1) and top 60% most relevant terms (34) were used for analysis. Further based upon Table 3.A1, Fig. 3.A1 shows bibliographic analysis in form of networking diagram. Further based upon Fig. 3.A1, research gap analysis with different nodes is shown in Fig. 3.A2. Tables 3.A2 and 3.A3, respectively, shows the significant work done on recycled PLA and year-wise publication details for future reference.

3.6 Summary

In the present study, feedstock filament of recycled PLA composite reinforced with Fe_3O_4 powder was developed on lab scale for three recycling stages. The targeted application for the developed feedstock was self-assembly characteristics using two-way programming method using magnetic field as an external stimulus. From the current research, following are the observations:

- From mechanical testing, it may be concluded that with the increasing recycling stages, the material strength of the reinforced matrix was seen to be decreased but the decrease in material performance was at low rate as only 22% decrement in mechanical strength was observed for the third stage of recycling.
- From VSM analysis, it may be concluded that the recycling stages have not affected the magnetic performance of material matrix as the reextrusion temperature was well below the curie temperature of the reinforced material.
- Further surface characteristics observed by Tool maker's microscope and 3D rendering software (3D rendered image and surface roughness profile) have given support to the observed mechanical trends.
- From statistical control testing for the process, it may be concluded that the magnetization of the PLA composite was stable and repeating over the recycling stages.
- The developed feedstock filament has shown repeatable two-way programming capability as the C_p and C_{pk} values of the magnetization were greater than 1.

Annex. 1 See Tables 3.A1–3.A3, and Figs. 3.A1 and 3.A2.

Table 3.A1 Relevance score with occurrences for keyword recycled PLA.

Id	Term	Occurrences	Relevance score
1	3D printing	6	2.2043
2	Bio	15	0.4619
3	Biodegradability	9	1.1752
4	Chain extender	7	1.1516
5	Characterization	22	0.7723
6	Chemical recycling	11	0.7115
7	Circular economy	7	0.6236
8	Comparison	6	1.3414
9	Degradation	30	0.3255
10	Depolymerization	6	0.7985
11	Design	13	1.1403
12	Development	11	0.9507
13	End	10	0.8691
14	Filament	14	2.176
15	Hydrolysis	10	0.5287
16	L lactide	12	0.7533
17	Lactide	12	0.6767
18	Life cycle assessment	12	0.8743
19	Life poly	6	1.0145
20	Mechanical	13	2.092
21	Mechanical properties	7	2.2907
22	Performance	13	0.7293
23	Pet	11	0.3261
24	Phospholipase	11	0
25	PLA composite	9	1.742
26	Plastic	21	0.3995
27	Polyester	10	0.6308
28	Polymer	12	0.4741
29	Preparation	16	1.1021
30	Product	9	0.9035
31	Production	21	0.6533
32	Properties	21	1.8908
33	Recycled poly	11	1.4266
34	Use	12	0.7898

Table 3.A2 Database of various countries engaged in recycled PLA (as per web of science database).

Countries/regions	Record count	Significant studies reported in past 20 years (% of 572)
People's Republic of China	98	17.133
United States	74	12.937
Spain	69	12.063
Italy	44	7.692
Germany	34	5.944
Japan	33	5.769
England	26	4.545
France	26	4.545
Thailand	25	4.371
India	24	4.196
Poland	24	4.196
Malaysia	20	3.497
Taiwan	20	3.497
Canada	19	3.322
Sweden	18	3.147
Brazil	17	2.972
The Netherlands	13	2.273
South Korea	12	2.098
Belgium	10	1.748
Turkey	9	1.573
Czech Republic	8	1.399
Iran	8	1.399
Portugal	8	1.399
Finland	7	1.224
Hungary	7	1.224

Showing 25 out of 68 entries

1 record(s) (0.175%) do(es) not contain data in the field being analyzed

Table 3.A3 Year-wise publications in web of science core collection.

Publication years	Record count	Significant studies reported in past 20 years (% of 572)
2019	89	15.559
2020	89	15.559
2018	70	12.238
2021	50	8.741
2014	41	7.168
2016	40	6.993
2015	34	5.944
2017	29	5.07
2009	20	3.497
2012	20	3.497
2011	17	2.972
2010	16	2.797
2013	15	2.622
2008	10	1.748
2006	7	1.224
2005	6	1.049
2007	5	0.874
2001	4	0.699
2003	3	0.524
1999	2	0.35
2000	2	0.35
2002	2	0.35
2004	1	0.175

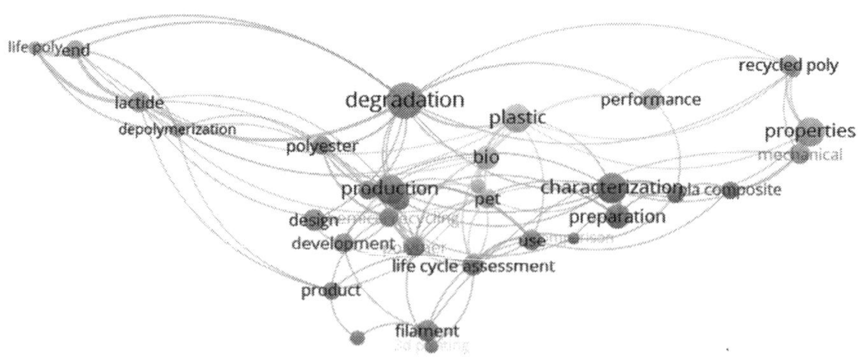

Figure 3.A1 Bibliographic analysis for keyword recycled PLA.

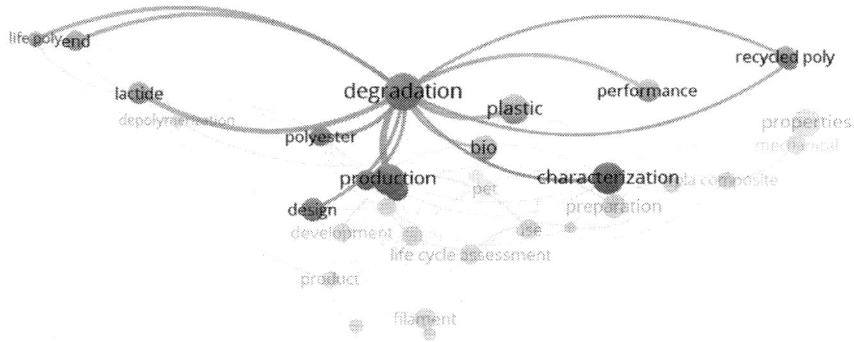

(A)

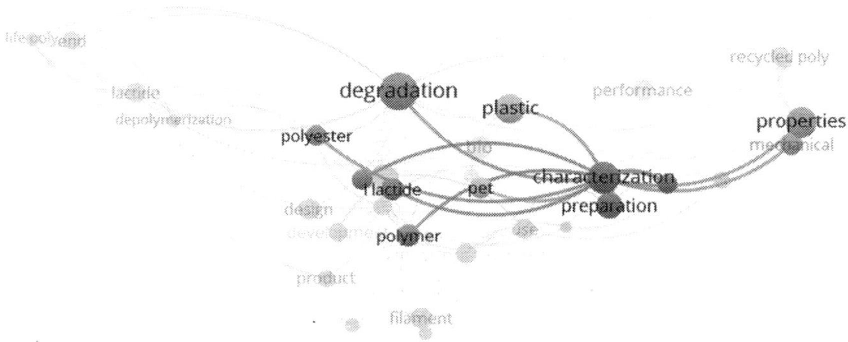

(B)

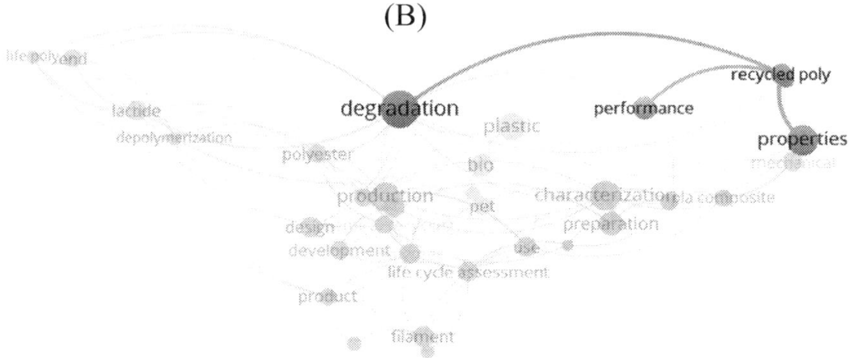

(C)

Figure 3.A2 Research gap analysis for keyword recycled PLA. (A) Bibliographic linkage for different thermoplastic for recycling. (B) Bibliographic linkage for thermoplastic with characterization and properties. (C) Biobliographic linkage for degrdadation, performance and recycled properties.

References

[1] Pei E. 4D printing: dawn of an emerging technology cycle. Assembly Automation 2014;34(4):310–14.
[2] Tibbits S. 4D printing: multi-material shape change. Architectural Design 2014;84 (1):116–21.
[3] Pei E. 4D printing—revolution or fad? Assembly Automation 2014;34(2):123–7.
[4] Dong Y, Toyao H, Itoh T. Compact circularly-polarized patch antenna loaded with metamaterial structures. IEEE Transactions on Antennas and Propagation 2011;59 (11):4329–33.
[5] Alves F, Grbovic D, Kearney B, Lavrik NV, Karunasiri G. Bi-material terahertz sensors using metamaterial structures. Optics Express 2013;21(11):13256–71.
[6] Ding J, Arigong B, Ren H, Zhou M, Shao J, Lu M, et al. Tuneable complementary metamaterial structures based on graphene for single and multiple transparency windows. Scientific Reports 2014;4(1):1–7.
[7] Lu M, Li W, Brown ER. Second-order bandpass terahertz filter achieved by multilayer complementary metamaterial structures. Optics Letters 2011;36(7):1071–3.
[8] Amorim DJ, Nachtigall T, Alonso MB. Exploring mechanical *meta*-material structures through personalised shoe sole design. In: Proceedings of the ACM symposium on computational fabrication, Jun 16, 2019. 1–8.
[9] Tibbits S. The emergence of "4D printing". In: TED Talk, Feb 2013.
[10] Kumar PK, Lagoudas DC. Introduction to shape memory alloys. Shape memory alloys. Boston, MA: Springer; 2008. p. 1–51.
[11] O'Handley Rorbert C. Model for strain and magnetization in magnetic shape-memory alloys. Journal of Applied Physics 1998;83(6):3263–70.
[12] Sun L, Huang WM. Nature of the multistage transformation in shape memory alloys upon heating. Metal Science and Heat Treatment 2009;51(11–12):573.
[13] Jani JM, Leary M, Subic A, Gibson MA. A review of shape memory alloy research, applications and opportunities. Materials & Design (1980–2015) 2014;56:1078–113.
[14] Kumar S, Singh R, Batish A, Singh TP. Additive manufacturing of smart materials exhibiting 4-D properties: a state of art review. Journal of Thermoplastic Composite Materials 2019. Available from: https://doi.org/10.1177/0892705719895052.
[15] Buehler WJ, Wang FE. A summary of recent research on the nitinol alloys and their potential application in ocean engineering. Ocean Engineering 1968;1(1):105–20.
[16] Zafar MQ, Zhao H. 4D printing: future insight in additive manufacturing. Metals and Materials International 2019;1–22.
[17] Tellinen J, Suorsa I, Jääskeläinen A, Aaltio I, Ullakko K. Basic properties of magnetic shape memory actuators. In: 8th international conference actuator, Jun 10, 2002. p. 566–569.
[18] Stoeckel D, Pelton A, Duerig T. Self-expanding nitinol stents: material and design considerations. European Radiology 2004;14(2):292–301.
[19] Mewissen MW. Self-expanding nitinol stents in the femoropopliteal segment: technique and mid-term results. Techniques in Vascular and Interventional Radiology 2004;7(1):2–5.

CHAPTER 4

3D printed graphene-reinforced polyvinylidene fluoride composite for piezoelectric properties

Vinay Kumar[1,2], Rupinder Singh[3] and I.P.S. Ahuja[2]
[1]Department of Production Engineering, Guru Nanak Dev Engineering College, Ludhiana, India
[2]Department of Mechanical Engineering, Punjabi University, Patiala, India
[3]Department of Mechanical Engineering, National Institute of Technical Teachers Training and Research, Chandigarh, India

4.1 Introduction

Fused deposition modeling (FDM) technique of additive manufacturing has gained so much popularity in last two decades because of great variation in quality of material, effectiveness of products manufactured, cost of production and durability of products in both engineering and medical applications. The invention of shape memory polymers, electroactive polymers, and one-way or two-way programmable polymers has changed the course of 3D printing of polymers toward 4D printing. Elastomer with composition of highly elastic polymers has been reported by some researchers to investigate the 4D printing applications of 3D printed prototype for self-folding applications [1]. Also, smart scaffolds with 4D properties have been reported for using shape memory polymers as biomedical implants [2]. A review of smart polymers and nanocomposites has shown that there are unending boundaries of researches related to capabilities of 4D printing applications like tissue engineering, bio-mimetic, civil infrastructure, origami, etc. [3,4]. Some recent studies conducted on 4D printing of electric stimulus controlled polylactic acid polymer matrix-based composite and thermal stimulus controlled methacrylate polymer has been reported that supports the revolution in manufacturing of industrial use products with intelligent characteristics. Polymers with thermal and electrical responsive properties can be very useful 3D/4D printing materials for development of self-active products [5,6]. Polyvinylidene fluoride

4D Printing
DOI: https://doi.org/10.1016/B978-0-12-823725-0.00009-6

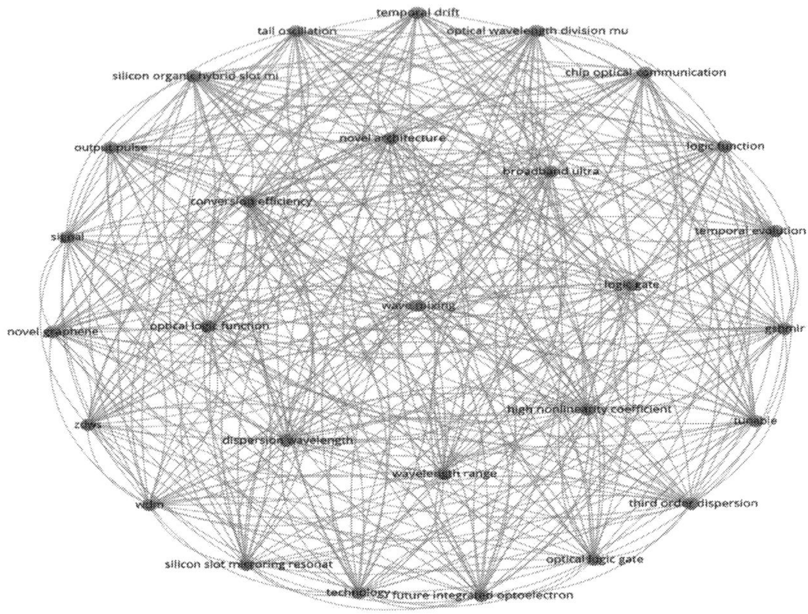

Figure 4.1 Web of key terms based on Scopus data base studied for 4D printing applications.

(PVDF) thermoplastic is one of the widely explored polymers for 3D printing-based 4D applications and electric conductivity-related properties that has been observed in its nanocomposite matrices [7–9]. Fig. 4.1 shows the web of keywords like tune-able polymers, energy storing semi-conductors and sensors, signal modems for variable wavelengths, etc. that have been investigated for 4D properties in various materials by the researchers around the globe.

Fig. 4.1 shows that a lot of studies have been performed on 4D properties of various materials and their combinations. Some researchers have reported 3D printed copolymer sensor, graphene-reinforced composite for power nanogenerators, smart terpolymer composites, and electroactive structural materials made using PVDF as base material [10–13]. Table 4.1 highlights the most relevant explored terms for characterization of 4D properties in various materials.

It has been observed that the most recently performed experimental investigations in which nanoparticles reinforcements (like graphene, Mn doped ZnO, and barium titanate) were used with polymer blends promoted the 4D characteristics in composites. Their research outcomes paved way for 4D printing applications of self-assembly, functionally

Table 4.1 List of terms and their respective relevance score to study 4D properties in materials.

ID	Term	Occurrences	Relevance score
1.	Actuator application	1	1.0422
2.	Case study	1	0.7574
3.	Clockwise relay operator	1	1.0422
4.	Conceptual design	1	0.7574
5.	Constant heat source	1	0.7574
6.	Electrical energy	1	0.7574
7.	Electrode system	1	1.4005
8.	Energy storage system	1	1.4005
9.	Enhanced super-capacitive performance	1	1.4005
10.	Feed-forward inverse compensation	1	1.0422
11.	Energy harvester	1	0.7574
12.	Harvester output characteristic	1	0.7574
13.	High specific capacitance	1	1.4005
14.	Hysteresis compensation	1	1.0422
15.	Hysteresis model	1	1.0422
16.	Hysteresis nonlinearity	1	1.0422
17.	Inverse compensator	1	1.0422
18.	Low cost chemical method	1	1.4005
19.	Mathematical model	1	0.7574
20.	Mechanical subsystem	1	0.7574
21.	Mesoporous structure	1	1.4005
22.	Model complexity	1	1.0422
23.	Model identification	1	1.0422
24.	Nano graphene	1	1.4005
25.	Nanostructure phases	2	0.8665
26.	Nanostructured material	1	1.4005
27.	Natural frequency	1	0.7574
28.	Numerical implementation	1	1.0422
29.	Numerical solution	1	0.7574
30.	Order and parameter of molecules	1	1.0422
31.	Physiochemical technique	1	1.4005
32.	Piezoelectric flexible cantilever beam	1	0.7574
33.	Reduced dimensional array	1	1.0422
34.	Renewable energy storage system	1	1.4005
35.	Selection mechanism	1	1.0422
36.	Shape memory alloy hysteretic behavior	1	0.7574
37.	Shape memory alloy wire	1	0.7574
38.	Simulation study	1	1.0422
39.	Smart technology device	1	1.4005
40.	Specific surface area	1	1.4005

(*Continued*)

Table 4.1 (Continued)

ID	Term	Occurrences	Relevance score
41.	Structure	5	0.802
42.	Superior electrode	1	1.4005
43.	Synthesized Zn powder	1	1.4005
44.	Temperature time behavior	1	0.7574
45.	Thermal energy harvester	1	0.7574
46.	Thermal subsystem	1	0.7574
47.	Vibrating action	1	0.7574

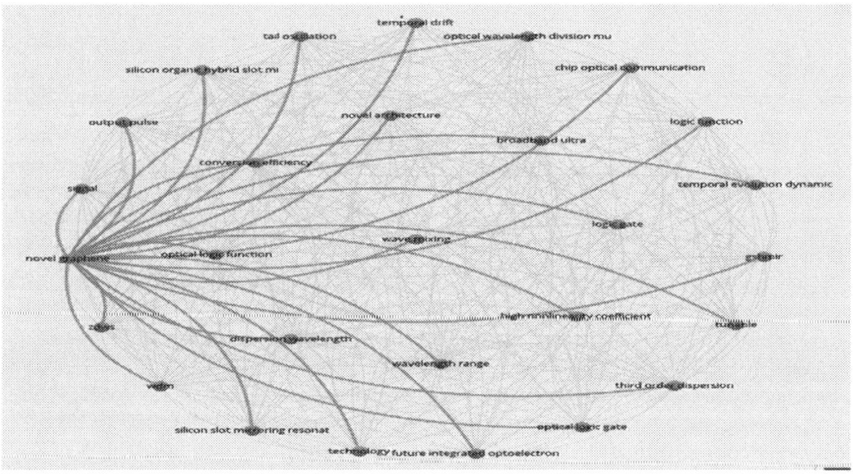

Figure 4.2 Research keywords linked to graphene reinforcement for 4D applications.

graded, and biocompatible materials with integration of 3D printing [14–21]. Fig. 4.2 shows various research terms linked to studies reported on graphene for 4D printing.

Therefore, some investigations based on chemical-assisted mechanical blending techniques are required to perform on PVDF like thermoplastic material in which there is a capability of phase changing (like α to β phase or vice versa with the help of stimulus) so that better results can be obtained in the area of 4D applications.

4.2 Research gap and problem formulation

The review of literature states that a very useful work has been reported on the applications of PVDF and its composite material for modern engineering and medical applications like fabrication of biomedical sensors,

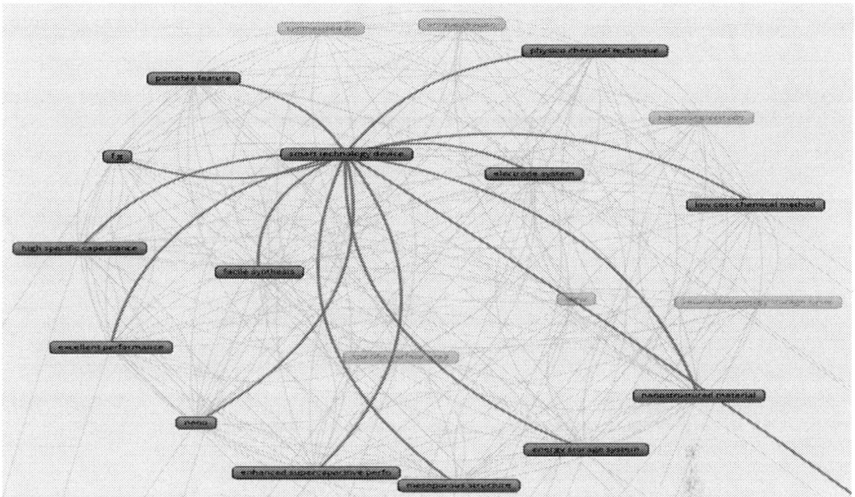

Figure 4.3 Research work reported on PVDF for nonconventional applications.

magnetic structures, nanocomposites type energy storage device, etc. Fig. 4.3 shows that properties of PVDF as a smart technological device, mesoporous structure, nanostructure material, high capacitance device have been explored by many researchers in the past. But hitherto a little is reported on investigating chemical blending of PVDF with highly conductive and considerable good in magnetic (like graphene) nanoparticles to prepare a smart composite with 4D properties to develop a low-cost and in-house manufactured material for 3D printing applications.

In the present work, PVDF and PVDF-graphene (PVDFG) composites were exposed to chemical-assisted mechanical blended (CAMB) so that a uniform blend of material can be obtained. The lumps of PVDF and PVDFG blends were analyzed for thermal stability and then processed to perform 3D printing of disk-shaped prototypes using open source FDM printer. Piezoelectric properties of PVDFG were investigated, and 4D capabilities were also ascertained by performing vibration sample magnetometry (VSM) test of the composites.

4.3 Experimentation

The study conducted in a previous work outlined that better rheological (melt flow index) and mechanical properties like peak strength (based on universal testing machine) can be imparted to PVDFG composites

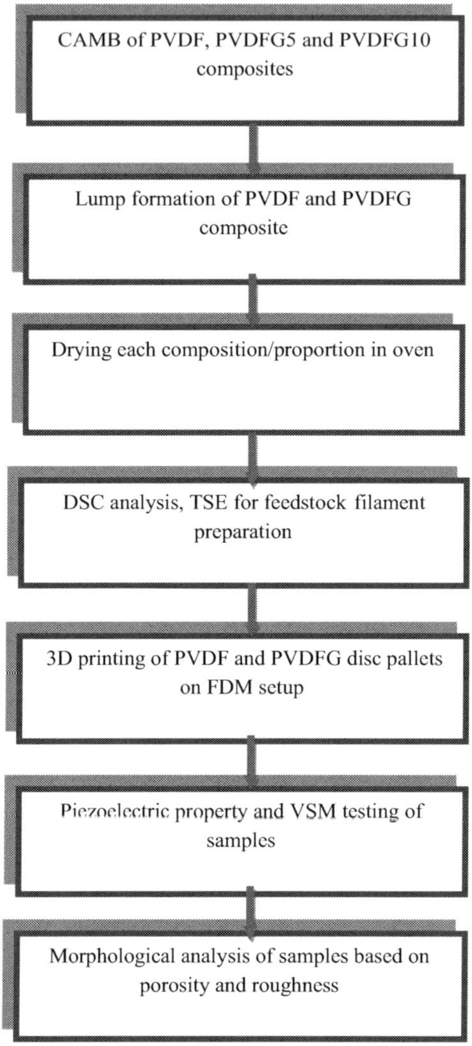

Figure 4.4 Work methodology for CAMB of PVDF and PVDFG composites 4D analysis.

prepared by mechanical blending process [14]. The work methodology for the present study of CAMB-PVDFG compositions/proportions as an extension of previous work [14] is listed in Fig. 4.4.

4.3.1 Chemical-assisted mechanical blended of PVDF-graphene composite

The PVDF polymer grains, grapheme nanoparticles, and N,N-dimethylformamide (DMF) chemical solvent were procured from the local market.

First PVDF was treated with DMF to ascertain the reactivity of chemical and PVDF polymer. Uniform stirring of PVDF slurry for 6 hours at 45°C resulted in loose lump formation of PVDF. The same procedure was followed for CAMB of PVDFG5 and PVDFG10 composites having 5% and 10% graphene by weight blended in PVDF polymer matrix. Fig. 4.5 shows the raw form of PVDF grains, nanoparticles of graphene used for CAMB and blend of PVDFG composite prepared in a glass beaker by CAMB in the presence of DMF solvent.

The PVDF and PVDFG compositions/proportions prepared were investigated for thermal characterization and stability by performing differential scanning calorimetric (DSC) test. Fig. 4.6 shows the DSC results of each composition/proportion. All the samples were allowed to heat and cool two cycles by varying the temperature from 30°C to 230°C. Each composition was found thermally stable, and also incremental change was observed in heat capacity PVDF and PVDFG with presence of graphene particles.

4.3.2 Twin screw extruder and 3D printing of PVDF-graphene

The loose lumps of PVDF and PVDFG composites were allowed to heat in vacuum oven for 2 hours at 40°C to obtain dry lumps of each composition. The lumps were then shredded to process the blends in twin screw extruder (TSE) shown in Fig. 4.7A. ThermoFisher scientific miniCTW TSE was used to prepare feedstock filament wire of PVDFG composite for FDM 3D printer shown in Fig. 4.7B and C, respectively. PVDFG composite was processed at 195°C in TSE under 0.2 Nm applied torque and 10 kg dead weight. The 3D printed component fabricated using PVDFG filament wire is also highlighted in Fig. 4.7C.

4.3.3 Piezoelectric testing

The 3D printed prototypes of PVDF, PVDFG5, and PVDFG10 were polished uniformly by using a rotary polishing cum grinding machine. The polished surfaces of all samples were poled on a 5 kV capacity DC poling unit for 3 hours, and piezoelectric property in terms of dielectric constant D_{33} was recorded. Fig. 4.8 shows (A) the setup for polishing samples, (B) 5 kV poling unit (Input type: DC), and (C) D_{33} meter.

4.3.4 Vibration sample magnetometry analysis

The magnetic properties of PVDFG5 and PVDFG10 were explored by VSM analysis. Being a conductive and magnetic material, graphene

Figure 4.5 PVDF (A), graphene particles (B), blend of PVDFG composite prepared by CAMB using DMF solvent (C).

particles induced magnetic field stimulus-based 4D properties in PVDFG composites. The magnetization and demagnetization in samples was recorded on application of 1Tesla applied magnetic field, that is, of the order 10^4 Gauss.

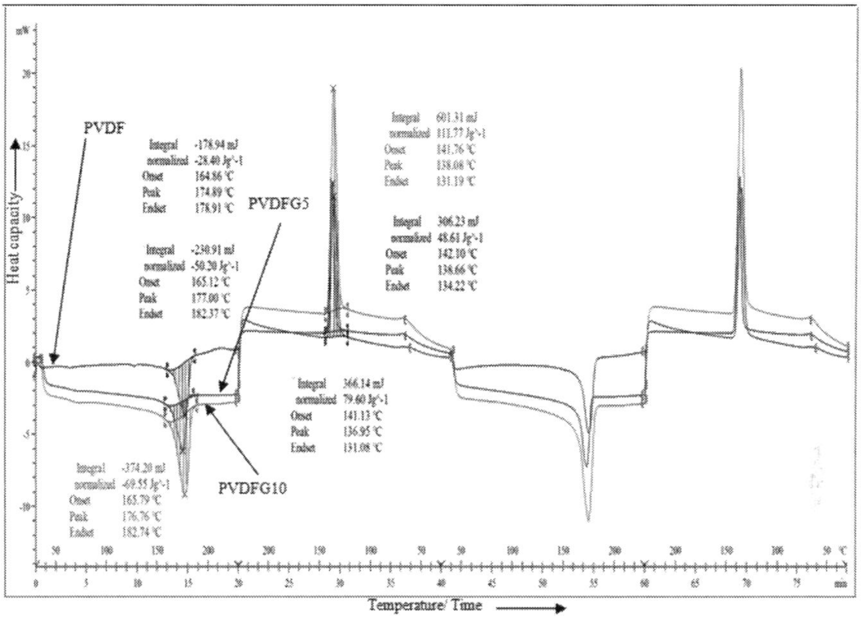

Figure 4.6 DSC results of PVDF and PVDFG composite.

4.4 Results and discussion

4.4.1 Thermal, piezoelectric, and vibration sample magnetometry analysis

The results obtained for thermal (DSC) testing, piezoelectric properties (DC poling and D_{33} effect), and VSM-based magnetic property is listed in Table 4.2.

The results in Table 4.2 shows that heat capacity of PVDF increases in a significantly positive manner without affecting the thermal stability of the base polymer matrix even after blending of nanoparticles of graphene. The composite PVDFG10 possessed highest heat capacity of 69.55 J/g as compared to PVDF that shown 28.40J/g heat capacity. PVDF sample was capable of DC poling at 1.5 kV for 2 hours. Beyond this limit, the PVDF grade used was not able to withstand at higher applied electric field as leakage of charge in the 3D printed sample takes place and setup shows the overloading signal. On the other hand, PVDFG composites were capable of more poling for longer time period. It can be observed that 3D printed PVDFG10 disk sample was able to pole up to 4.5 kV for 3 hours.

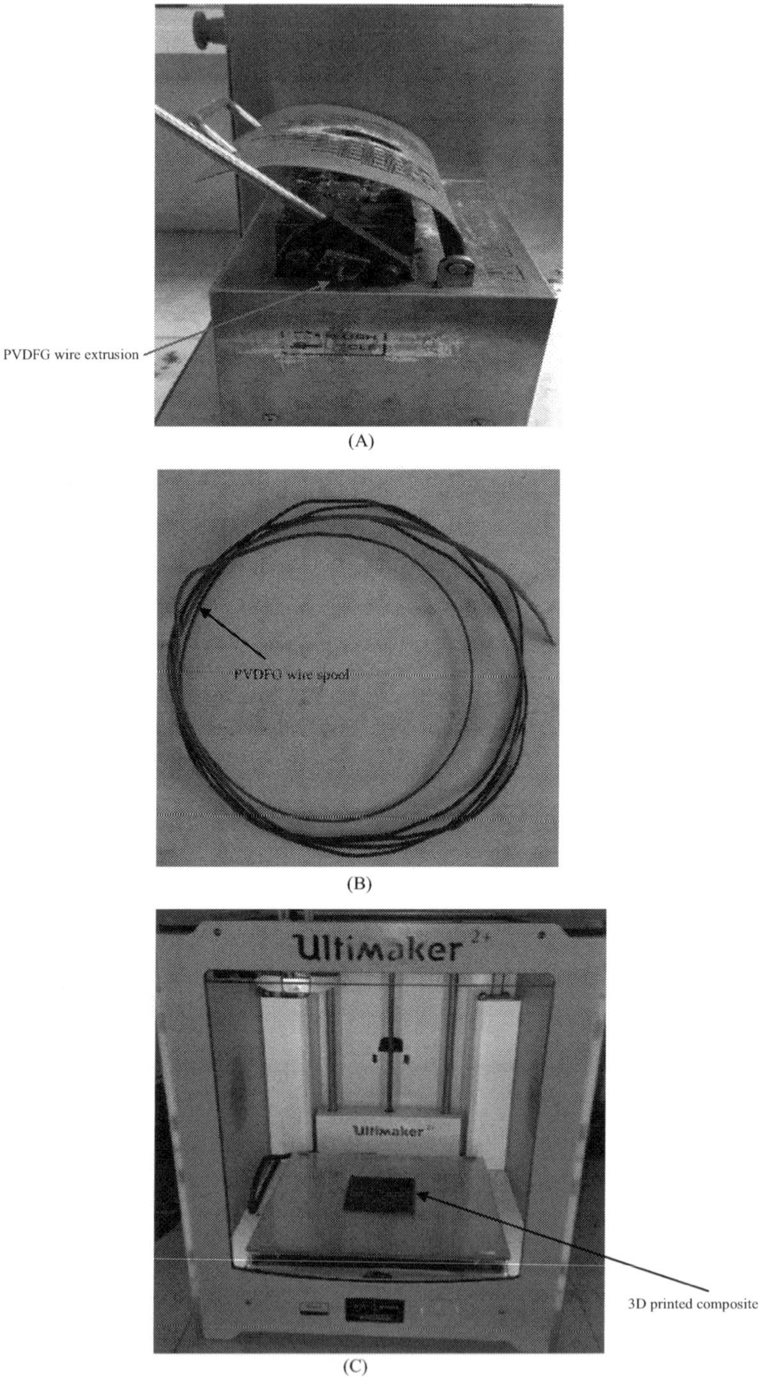

Figure 4.7 Twin screw extruder (A), PVDFG feedstock filament wire (B), and FDM-based 3D printer (C).

With increase in weight percentage of graphene nanoparticles in PVDF, piezoelectric properties of polymer matrix increased appreciably. Also, D_{33} of PVDF improved as PVDFG10 composite shows 47.6pC/N D_{33} value was sufficiently much greater than 0.3pC/N D_{33} of PVDF. Physical significance of piezoelectric property can be understood with the schematic shown in Fig. 4.9.

Surface of 3D printed disk samples got charged by the application of external electric field and then very high frequency load is applied on the specimens to observe 3 directional volumetric changes in terms of D_{33}. It can be stated that electrical conductivity can be increased in PVDF by chemically blending the highly conductive graphene particles. The electric charge acts as a stimulus to induce β phase in the composite and application of externally applied load produced 4D properties in PVDFG composite. The PVDFG composites were also found magnetic responsive as the magnetization graph shown in Fig. 4.10 highlighted a significant magnetic response of PVDFG specimens. PVDFG10 was best acceptable composition with magnetization of 0.0853×10^{-5} emu/g. The proposed composition/proportion of PVDFG can be used as a 4D magnetic sensor and actuator material that possesses self-active properties. Such composites can be explored for self-contracting and self-expanding materials for repairing and filling gaps in the civil structures by giving the input of electric field.

4.4.2 Morphological analysis

The Tool Maker's microscopy-based morphological results obtained for PVDF and PVDFG composites are shown in Fig. 4.11. Fig. 4.11A–C, respectively, shows cross-sectional view of PVDF, PVDFG5, and PVDGG10 wires along radial axis (as per Table 4.2) at ×100 magnification. The porosity percentage was calculated using metallurgical image analysis software as per ASTM B 276 (Fig. 4.11D–F). High porosity percentage of 21.04% was observed in PVDF (without reinforcement). The morphological properties were improved for PVDFG5 and PVDFG10 for which porosity percentage was noticed as 14.56% and 9.79%, respectively. Further open source image processing tool was used to get 3D rendered image (Fig. 4.11G–I) corresponding to Fig. 4.11A–C. Fig. 4.11J–L shows the pictorial representation for surface texture, waviness, and roughness plots. Corresponding surface roughness values at cutoff length of 0.05 mm is shown in Fig. 4.11M–O. As observed from Fig. 4.11M–O, the surface roughness was observed as 89.2 nm, 51.39 nm, and 50.44 nm, which

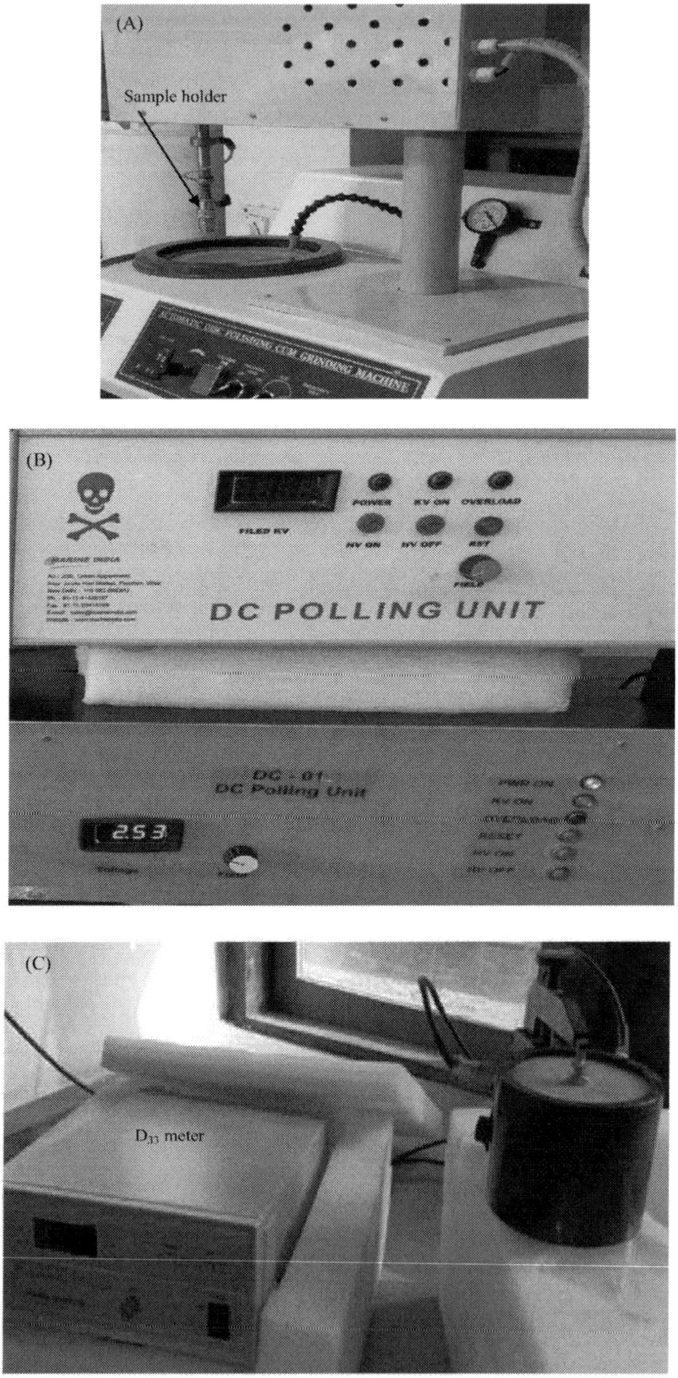

Figure 4.8 Pneumatic disk polishing setup (A), 5 kV DC poling unit (B), D_{33} measuring instrument (C).

Table 4.2 Results obtained for 3D/4D properties of PVDF and PVDFG composites.

S. No.	Composition/ proportion	Heat capacity (J/g)	DC poling limit (kV) & duration	D_{33} (pC/N)	Magnetization (emu/g)
1.	PVDF	28.40	1.5 kV for 2 hours	0.3	0
2.	PVDFG5	50.20	2.53 kV for 3 hours	14.2	0.0216×10^{-5}
3.	PVDFG10	69.55	4.5 kV for 3 hours	47.6	0.0853×10^{-5}

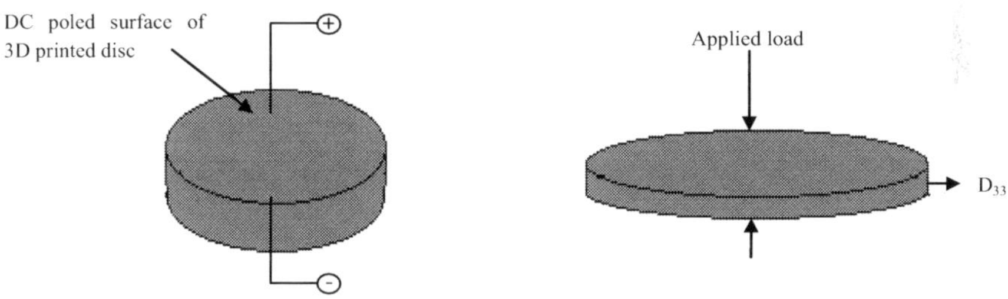

Figure 4.9 Schematics of piezoelectric properties in 3D printed PVDF/PVDFG composites.

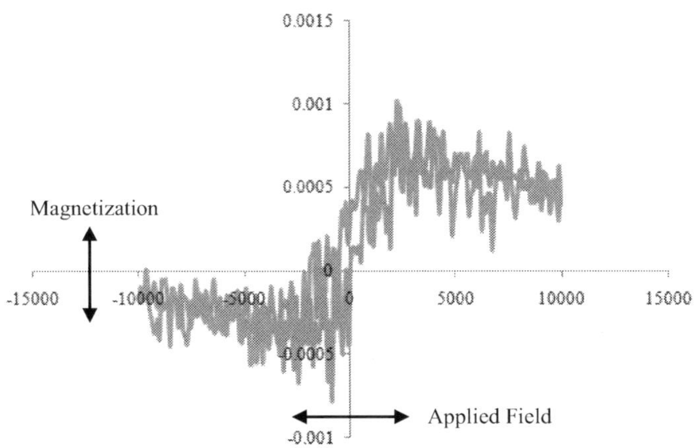

Figure 4.10 VSM graph for PVDFG composite.

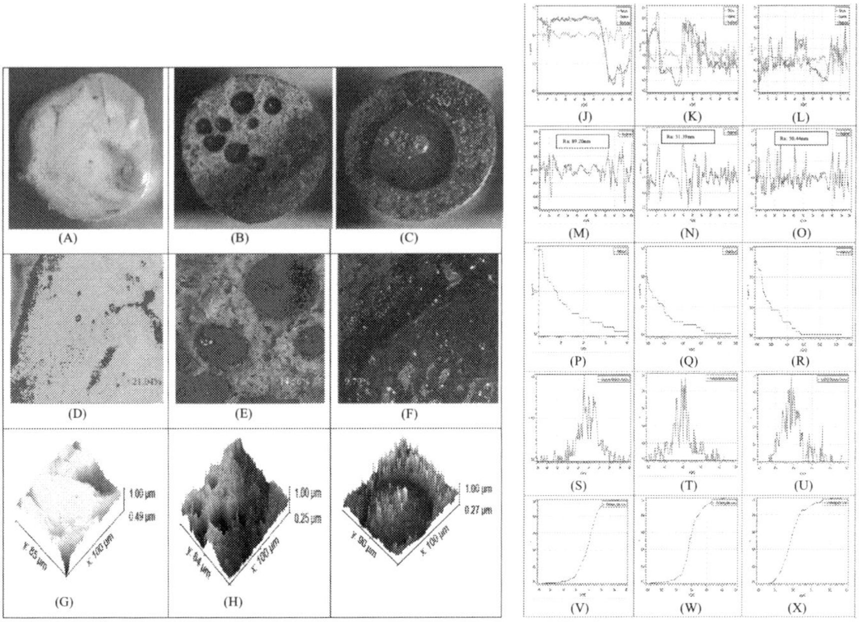

Figure 4.11 Tool Maker's microscopy and porosity % images for PVDF, PVDFG5, and PVDFG10 composite wire samples along radial axis. Cross-sectional view of wire samples along radial axis: PVDF (A), PVDFG5 (B) and PVDGG10 (C), porosity percentage: PVDF (D), PVDFG5 (E) and PVDGG10 (F), 3D rendered images: PVDF (G), PVDFG5 (H) and PVDGG10 (I), surface texture, waviness and roughness plots: PVDF (J), PVDFG5 (K) and PVDGG10 (L), surface roughness values: PVDF (M), PVDFG5 (N) and PVDGG10 (O), peak count plots: PVDF (P), PVDFG5 (Q) and PVDGG10 (R), amplitude distribution function plots: PVDF (S), PVDFG5 (T) and PVDGG10 (U), bearing ratio curve plots: PVDF (V), PVDFG5 (W) and PVDGG10 (X).

justifies better surface properties for PVDFG10 composite. Further peak count (Fig. 4.11P–R), amplitude distribution function (Fig. 4.11S–U), and bearing ratio curves (Fig. 4.11V–X) were plotted for PVDF, PVDFG5, and PVDFG10 composites, which is in-line with the surface roughness observations.

4.5 Summary

- It may be concluded from the present study that PVDFG composite can be considered as a useful 3D printable smart material for 4D applications as PVDFG10 has shown acceptable thermal stability in terms of heat capacity (69.55J/g) after CAMB process to fabricate feedstock

filament for 3D printer. Acceptable piezoelectric properties were obtained in PVDFG composites that can be very useful to increase the industrial applications of PVDF for manufacturing of 3D printed sensors and actuators with 4D capabilities;
- The D_{33} of 47.6pC/N was obtained in PDVFG10 composite which shows that significant physical change can be induced in the proposed composite by application of electric discharge in it;
- The results of VSM test show that PVDFG10 can be treated as a useful magnetic sensor application material for robotics. The composite can be magnetized up to the range 0.0853×10^{-5} emu/g by applying 1Tesla magnetic field. The demagnetization capability of PVDFG shows 4D behavior in the composite;
- The morphological results also support the research outcomes as very less porosity percentage was observed in PVDFG10 (9.79%) composite which states that better 3D printed composite can be fabricated for exploring 4D applications of the proposed material.

The 3D printed PVDFG composites with 4D properties may be used to develop customized solution with self-contracting and self-expanding properties for self-healing of cracks developed in the structures raised in past centuries. The compatibility of customized solution can be achieved by reinforcement or blending of debris in the proposed solution to increase the binding capability of the base material and 3D printed composite material.

Acknowledgments

The authors are thankful to Department of Science and Technology (DST)-GoI for the financial support under research project File No. DST/TDT/SHRI-35/2018 and Manufacturing Research Lab, GNDEC Ludhiana for providing lab facilities and support.

References

[1] Yamamura S, Iwase E. Hybrid hinge structure with elastic hinge on self-folding of 4D printing using a fused deposition modeling 3D printer. Materials & Design 2021;109605.

[2] Zhang C, Cai D, Liao P, Su JW, Deng H, Vardhanabhuti B, et al. 4D Printing of shape-memory polymeric scaffolds for adaptive biomedical implantation. Acta Biomaterialia 2021;122:101−10.

[3] Falahati M, Ahmadvand P, Safaee S, Chang YC, Lyu Z, Chen R, et al. Smart polymers and nanocomposites for 3D and 4D printing. Materials Today 2020.

[4] Subash A, Kandasubramanian B. 4D printing of shape memory polymers. European Polymer Journal 2020;109771.

[5] Shao LH, Zhao B, Zhang Q, Xing Y, Zhang K. 4D printing composite with electrically controlled local deformation. Extreme Mechanics Letters 2020;39:100793.
[6] Han M, Yang Y, Li L. Energy consumption modeling of 4D printing thermal-responsive polymers with integrated compositional design for material. Additive Manufacturing 2020;34:101223.
[7] Kumar R, Singh R, Singh M, Kumar P. On ZnO nano particle reinforced PVDF composite materials for 3D printing of biomedical sensors. Journal of Manufacturing Processes 2020;60:268–82.
[8] Kumar R, Kumar P. Co doped ZnO nanoparticles reinforcements in PVDF for 3D printing of magnetic structures. Reference module in materials science and materials engineering. Elsevier; 2020.
[9] Kim H, Johnson J, Chavez LA, Rosales CAG, Tseng TLB, Lin Y. Enhanced dielectric properties of three phase dielectric MWCNTs/BaTiO3/PVDF nanocomposites for energy storage using fused deposition modeling 3D printing. Ceramics International 2018;44(8):9037–44.
[10] Marandi M, Tarbutton J. Additive manufacturing of single-and double-layer piezoelectric PVDF-TrFE copolymer sensors. Procedia Manufacturing 2019;34:666–71.
[11] Badatya S, Kumar A, Sharma C, Srivastava AK, Chaurasia JP, Gupta MK. Transparent flexible graphene quantum dot-(PVDF-HFP) piezoelectric nanogenerator. Materials Letters 2021;129493.
[12] Tuluk A, Mahon T, van der Zwaag S, Groen P. Estimating the true piezoelectric properties of BiFeO3 from measurements on BiFeO3-PVDF terpolymer composites. Journal of Alloys and Compounds 2021;159186.
[13] Ribeiro C, Costa CM, Correia DM, Nunes-Pereira J, Oliveira J, Martins P, et al. Electroactive poly (vinylidene fluoride)-based structures for advanced applications. Nature protocols 2018;13(4):681.
[14] Kumar V, Singh R, Ahuja IPS, et al. On nanographene-reinforced polyvinylidene fluoride composite matrix for 4D applications. Journal of Materials Engineering and Performance 2021. Available from: https://doi.org/10.1007/s11665-021-05459-z.
[15] Kumar V, Singh R, Ahuja IPS. Secondary recycled acrylonitrile–butadiene–styrene and graphene composite for 3D/4D applications: rheological, thermal, magnetometric, and mechanical analyses. Journal of Thermoplastic Composite Materials 2020. p. 0892705720925114.
[16] Kumar R, Singh R, Kumar V, Kumar P. On Mn doped ZnO nano particles reinforced in PVDF matrix for fused filament fabrication: mechanical, thermal, morphological and 4D properties. Journal of Manufacturing Processes 2021;62:817–32.
[17] Kumar S, Singh R, Singh TP, Batish A. Investigations for magnetic properties of PLA-PVC-Fe3O4-wood dust blend for self-assembly applications. Journal of Thermoplastic Composite Materials 2019. p. 0892705719857778.
[18] Kumar S, Singh R, Singh TP, Batish A. Multimaterial printing and characterization for mechanical and surface properties of functionally graded prototype. Proceedings of the Institution of Mechanical Engineers, Part C: Journal of Mechanical Engineering Science 2019;233(19–20):6741–53.
[19] Sharma R, Singh R, Batish A. Study on barium titanate and graphene reinforced PVDF matrix for 4D applications. Journal of Thermoplastic Composite Materials 2019. p. 0892705719865004.
[20] Sharma R, Singh R, Batish A. On effect of chemical-assisted mechanical blending of barium titanate and graphene in PVDF for 3D printing applications. Journal of Thermoplastic Composite Materials 2020. p. 0892705720945377.
[21] Ranjan N, Singh R, Ahuja IPS. Development of PLA-HAp-CS-based biocompatible functional prototype: a case study. Journal of Thermoplastic Composite Materials 2020;33(3):305–23.

CHAPTER 5

On characterization of rechargeable, flexible electrochemical energy storage device

Kamaljit Singh Boparai[1], Abhishek Kumar[1] and Rupinder Singh[2]
[1]Department of Mechanical Engineering, GZSCCET, MRS Punjab Technical University, Bathinda, India
[2]Department of Mechanical Engineering, National Institute of Technical Teachers Training and Research, Chandigarh, India

5.1 Introduction

Due to rapid advancement in modern technologies such as electric vehicles, smart houses, bio-medical devices, and other hybrid energy storage technologies, demand for electrochemical energy is exceptionally increased which cannot be overlooked [1,2]. To keep up with the technological advancements and fulfill the demand of insufficient supply, many industries and researchers started exploring and catering interests toward developing more thermal stable, economical, efficient, and eco-friendly novel materials or improving existing materials to utilize their full potential for electrochemical energy storage and conversion devices [3−5]. Due to constant progress in hardware, functionality, software, and materials in fused deposition modeling (FDM) process widened the scope of printing complex and functional 3D geometries [6−8]. Previously, a large amount of research work is carried out and centered toward building parts of electrochemical energy storage devices such as electrodes, current collectors, separators, etc. [9−12]. Moreover, 3D printing (3DP) is also considered as advanced technique for building electrochemical energy storage devices because of the ability to seamlessly integrate into product structure, flexibility, and printing complex geometries.

FDM is a rapid prototyping technique that belongs to the metal extrusion family of additive manufacturing (AM) or 3DP [13]. AM is a material addition manufacturing process in which material deposited layer-by-layer according to predetermined path and finally fused in single form [14].

4D Printing
DOI: https://doi.org/10.1016/B978-0-12-823725-0.00007-2

In FDM, computer aided draft based 3D model in ".stl" or ".amf" extensions are converted into "G-code extension" using slicing software's like Cura, Simplify3D, Creality slicer, etc. The properties of parts produced by FDM depend upon the material that is being used as feedstock filament.

It has been reported that phase change materials have high thermal energy storage capabilities with a moderate temperature variation. Due to these properties, these materials have been popular among researches involving energy storage materials.

For ascertain the research gap in the proposed area, web of science database of past 20 years was explored (2001–20). Initially the search was performed with keyword "electrochemical energy storage device" and 11,965 results were obtained from core collection of web of science database. Out of these results, recent 500 articles were selected for further analysis. The VOS viewer (open-source software) was used for bibliographic analysis and preparation of networking diagrams. For the selected keyword by fixing minimum number of occurrences as "5", out of 12,005 terms, 667 meet the threshold. For each of 667 terms, relevance score was calculated and further 60% most relevant terms were used for analysis. Fig. 5.1 shows networking diagram for keyword electrochemical energy storage device. As observed from Fig. 5.1, 04 clusters were noticed

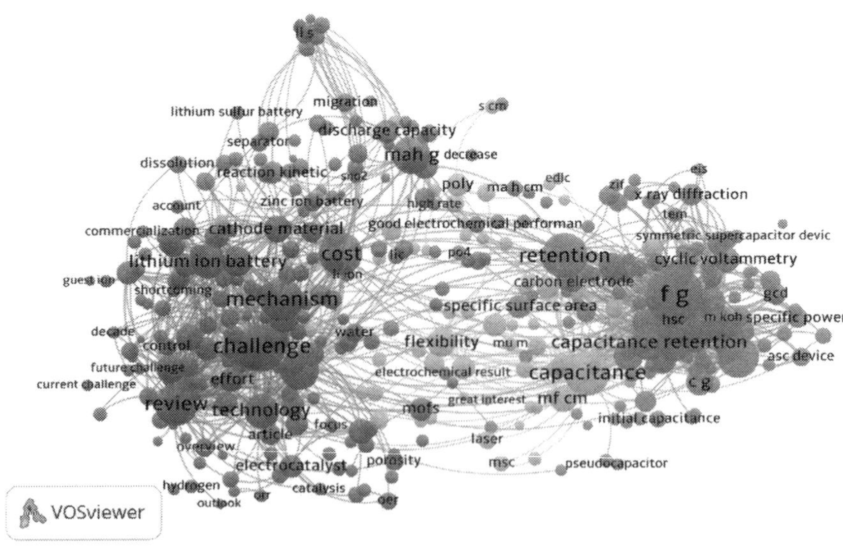

Figure 5.1 Networking diagram for keyword electrochemical energy storage device.

which represents the major work reported in particular application domain.

Further to ascertain the research gap, Fig. 5.2A–C shows research gap analysis by selecting nodal points as cost, specific capacitance, and mechanism. After this, the web of science database was explored for keyword "3D printed energy storage device" and 209 results were obtained. For the selected keyword by fixing minimum number of occurrences as "5", out of 6101 terms, 260 meet the threshold. For each of 260 terms, relevance score was calculated and further 60% most relevant terms (156) were used for analysis. Fig. 5.3 shows networking diagram for keyword 3D printed energy storage device. As observed from Fig. 5.3, 04 clusters were noticed which represents the major work reported in particular application domain. Further to ascertain the research gap, Fig. 5.4A–C shows research gap analysis by selecting nodal points as storage device, energy density, and flexible energy storage device. Finally, the keyword "rechargeable, flexible electrochemical energy storage device" was used, and total 125 results were obtained in core selection of web of science data. For the selected keyword by fixing minimum number of occurrences as "5", out of 3637 terms, 144 meet the threshold. For each of 144 terms, relevance score was calculated and further 60% most relevant terms (86) were used for analysis (Table 5.1). Fig. 5.5 shows networking diagram for keyword "rechargeable, flexible electrochemical energy storage device." As observed from Fig. 5.5, 03 clusters were noticed which represents the major work reported in particular application domain. Further to ascertain the research gap, Fig. 5.6A–C shows research gap analysis by selecting nodal points as flexible device, flexible battery, and high reversible capacity.

The web of science-based research gap analysis highlighted that the limited work has been done on the production of energy storage device with 3DP. The purpose of this research work is to characterize and validate new approach to fabricate low-cost/in-house composite feedstock filament material for FDM, which can be selected for the direct 3DP of energy storage devices. However, the selection of materials, that is (paraffin wax, graphite and graphene), has been done based upon the electrochemical storage application. The feedstock filament was fabricated on TSE after assessment of its rheological behavior by Melt Flow Index (MFI) tester. Standard parameters in software were specified like rotating screw speed, roller speed, and barrel temperature for the successful extrusion of feedstock filament. As a result, prepared compositions used for

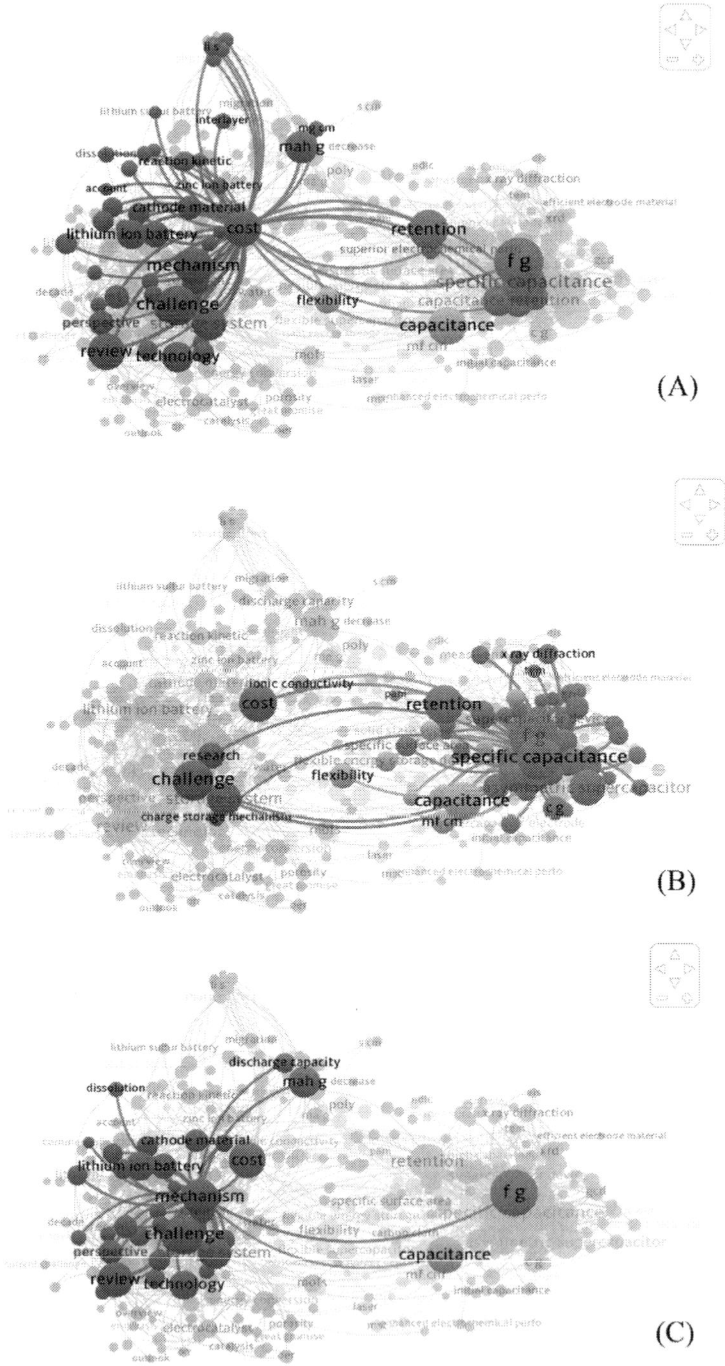

Figure 5.2 Gap analysis by selecting nodal points as cost (A), specific capacitance (B), and mechanism (C).

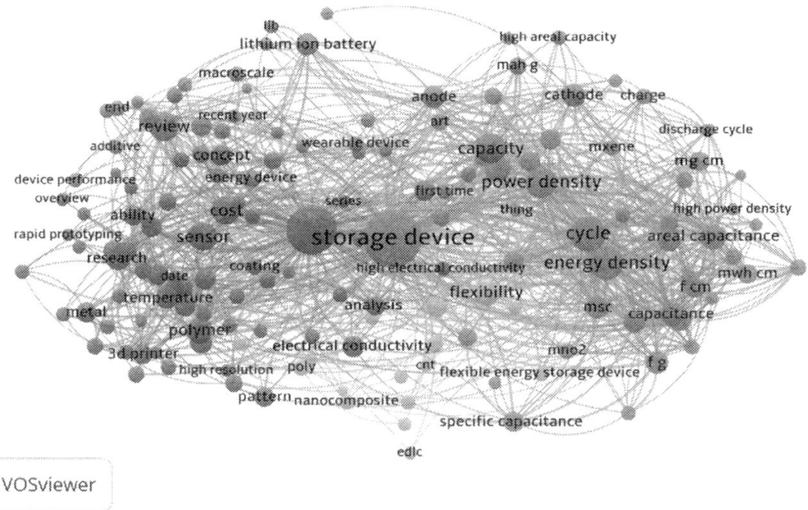

Figure 5.3 Networking diagram for keyword "3D printed energy storage device".

fabrication of feedstock filament for FDM applications were successfully developed and examined for its processability. Further, the structural morphology was realized by using scanning electron microscopy (SEM) and Fourier-transform infrared spectroscopy (FT-IR) analysis. Finally, the thermal properties and thermal stability of proposed materials were measured by using differential scanning calorimetry (DSC) to reveal its suitability as a feedstock filament (other than standard filament) for FDM and to cater various energy storage applications. The following methodology has been used for this research work. Fig. 5.7 shows the process flow diagram.

5.2 Experimentation

5.2.1 Material selection

The following materials are selected for extrusion applications:

Paraffin wax, chemical formula "$C_nH_{2n} + 2$," (Purity: >99%, density: 0.9 g/cm^3 melting point 40°C and 0.50% oil content) was supplied from Meta wares India Private limited, New Delhi. Paraffin waxes are a mixture of saturated hydrocarbon and normally consist of straight chain n-alkanes, $CH^3{-}(CH^2)_n{-}CH^3$. Moreover, the crystallization of the $(CH^2)_n$ chain releases a large amount of latent heat. Paraffin wax (Fig. 5.8A) is a most suitable material for heat storage, having a specific heat capacity of

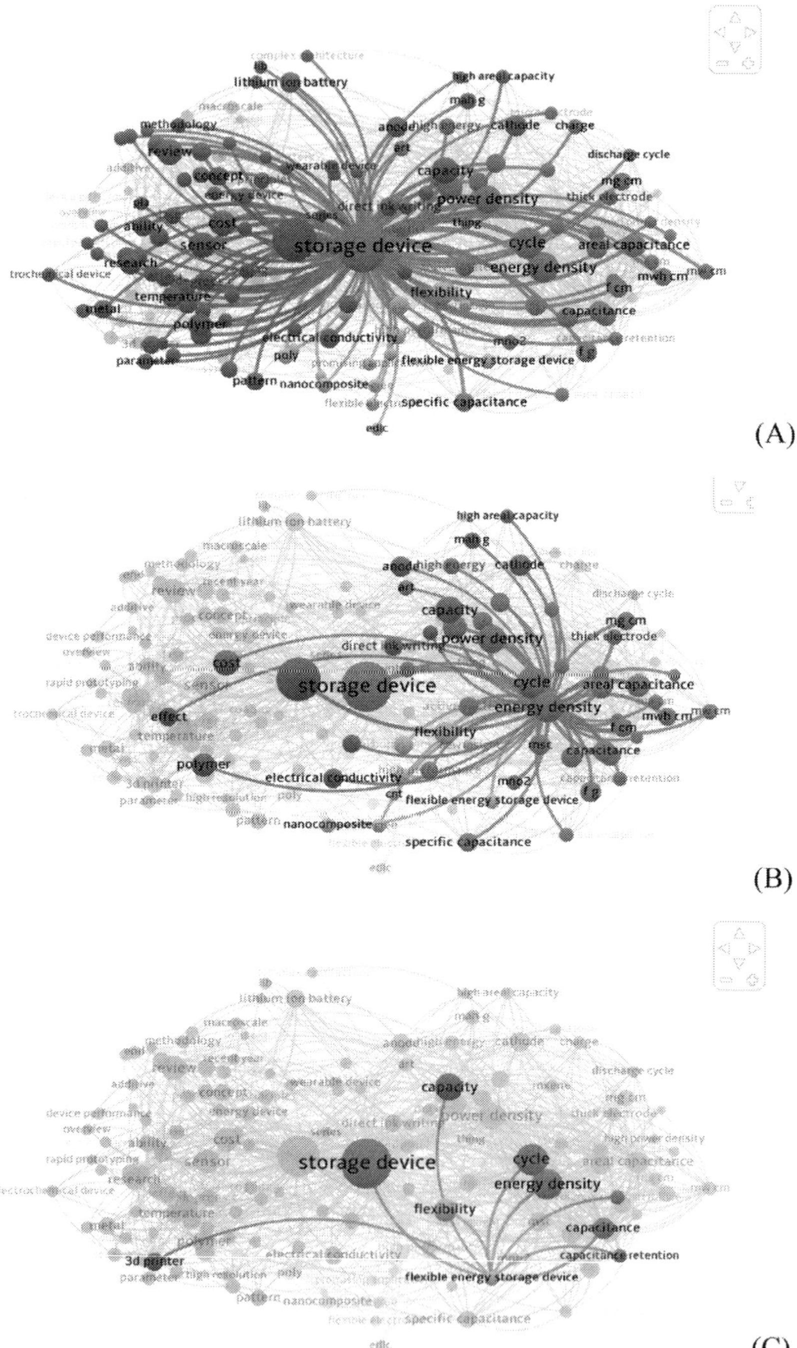

Figure 5.4 Gap analysis by selecting nodal points as storage device (A), energy density (B), and flexible energy storage device (C).

Table 5.1 Relevance score calculation for keyword rechargeable, flexible electrochemical energy storage device.

Id	Term	Occurrences	Relevance score
1	Advantage	14	0.6288
2	Capacity retention	20	1.5682
3	Carbon	9	0.2628
4	Carbon cloth	7	0.8798
5	Carbon nanotube	13	0.4817
6	Cathode material	12	0.5597
7	Challenge	29	0.9342
8	Characterization	5	0.2985
9	Cnt	7	0.7666
10	Cost	30	0.0674
11	Current collector	14	0.5594
12	Current density	13	1.4268
13	Cycle	48	0.9588
14	Cycling	8	1.0011
15	Cycling stability	19	0.9574
16	Degrees C	6	0.4469
17	Electric vehicle	14	0.5227
18	Excellent electrochemical performance	8	2.2603
19	Exploration	6	0.608
20	Fiber	14	1.0931
21	Field	15	1.0286
22	First time	7	1.0012
23	Flexible battery	9	0.6598
24	Flexible device	7	0.4313
25	Flexible electrode	8	0.338
26	Flexible electronic device	8	0.6045
27	Focus	6	1.6639
28	Great challenge	5	0.9335
29	High capacity	11	0.853
30	High electrochemical performance	5	0.7844
31	High energy	9	0.3233
32	High flexibility	5	0.5775
33	High performance	9	1.0109
34	High power density	6	0.6482
35	High reversible capacity	7	1.7545
36	High safety	9	2.1958
37	High specific capacity	7	0.6717
38	Hydrogel electrolyte	7	1.0214
39	Insight	7	1.632
40	Integration	9	0.4122

(*Continued*)

Table 5.1 (Continued)

Id	Term	Occurrences	Relevance score
41	Layer	12	0.6353
42	Li-ion battery	8	0.5497
43	Lib	7	0.609
44	Libs	10	0.7146
45	Light	6	0.8737
46	Limitation	7	0.585
47	Lithium	9	1.0812
48	Lithium-ion battery	22	0.7377
49	Long cycle life	7	1.2905
50	Ma cm	5	3.5486
51	Ma g	10	1.3164
52	Ma h cm	6	2.018
53	Ma h g	20	1.0732
54	Mah g	16	0.7707
55	Metal	8	0.4093
56	Miniaturization	5	1.1216
57	Morphology	5	0.3308
58	Mw cm	6	2.7879
59	Mw h cm	7	2.2398
60	Mwh cm	5	1.872
61	Nanomaterial	8	0.5221
62	Order	5	1.2232
63	Overview	6	1.5671
64	Paper	15	0.3892
65	Perspective	16	1.522
66	Poly	7	0.9958
67	Polymer	11	0.3899
68	Polymer electrolyte	9	0.7754
69	Positive electrode	5	2.245
70	Practical application	9	1.1346
71	Preparation	7	1.2294
72	Promising candidate	7	1.5011
73	Proof	6	1.7071
74	Recent advance	11	1.3578
75	Recent progress	7	1.5299
76	Recent year	9	0.9297
77	Research	17	0.5784
78	Review	22	1.1957
79	Self	11	0.5437
80	State	19	0.4855
81	Technology	25	1.0743

(Continued)

Table 5.1 (Continued)

Id	Term	Occurrences	Relevance score
82	Time	13	0.8878
83	Today	5	1.1603
84	Use	15	0.7534
85	Wearable electronic device	8	0.2857
86	Zinc-ion battery	8	0.6232

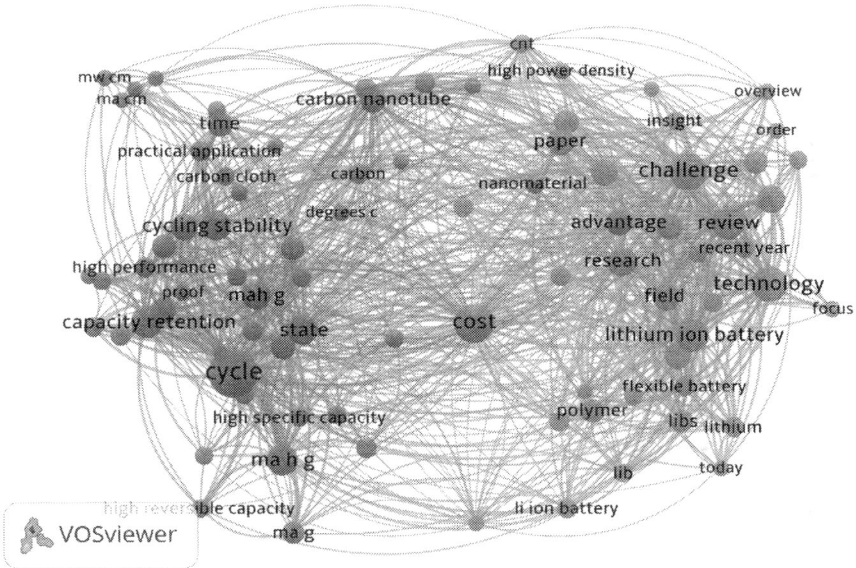

Figure 5.5 Networking diagram for keyword rechargeable, flexible electrochemical energy storage device.

2.14–2.9 J/g/K^{-1} (joules per gram kelvin) and a heat of fusion of 200–220 J/g. Graphite powder (Fig. 5.8B) with a purity of 99%, density 2.26 g/cm^{3}and particle size 200 mesh was supplied from Akshar Chem Solutions, India because of its favorable thermal conducting properties, it is an allotropic form of carbon, which exhibits two crystalline structure. Graphene powder (Fig. 5.8C) was supplied from Platonic Nanotech Private Limited having Purity: >99%, density: 0.15 g/cm^3, thickness of 5–10 nm, length 5–10-micron, average No. of layer 4–8, and surface area 190 m^2/g. The selection of this material is due to its promising thermal, mechanical, and chemical properties.

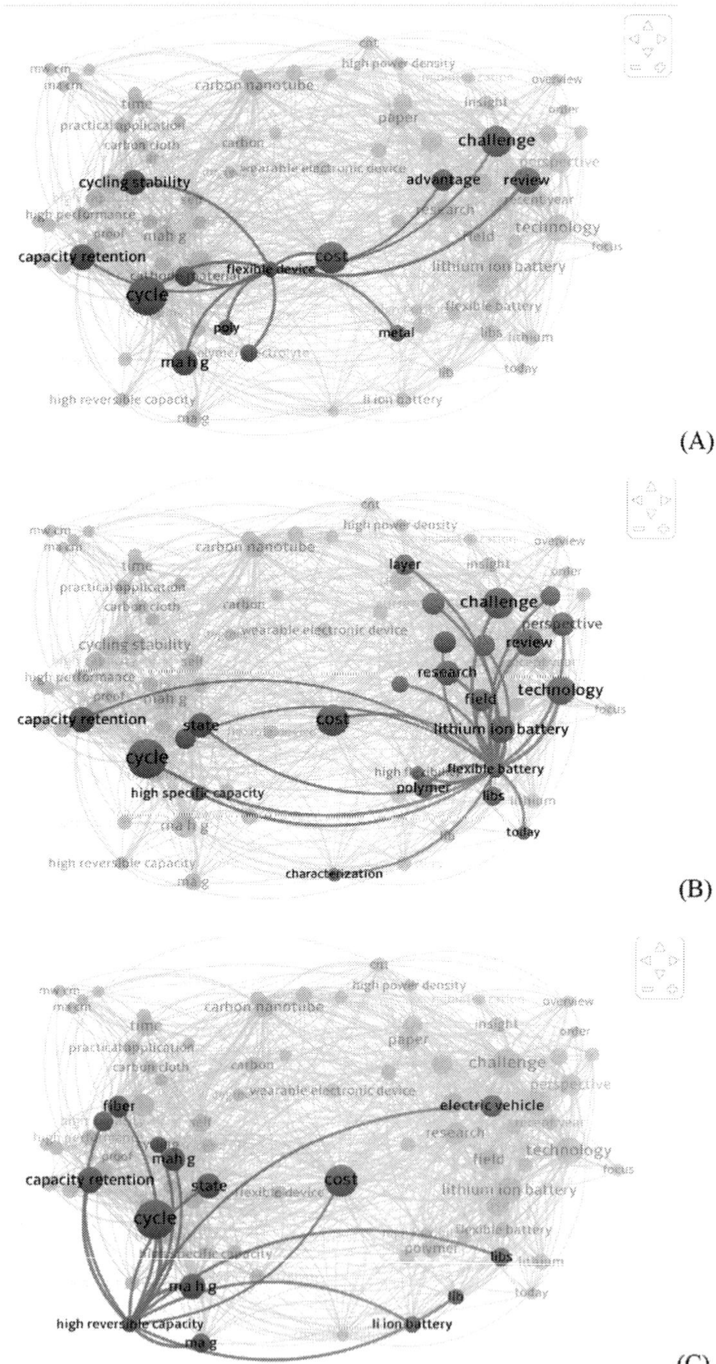

Figure 5.6 Gap analysis by selecting nodal points as flexible device (A), flexible battery (B), and high reversible capacity (C).

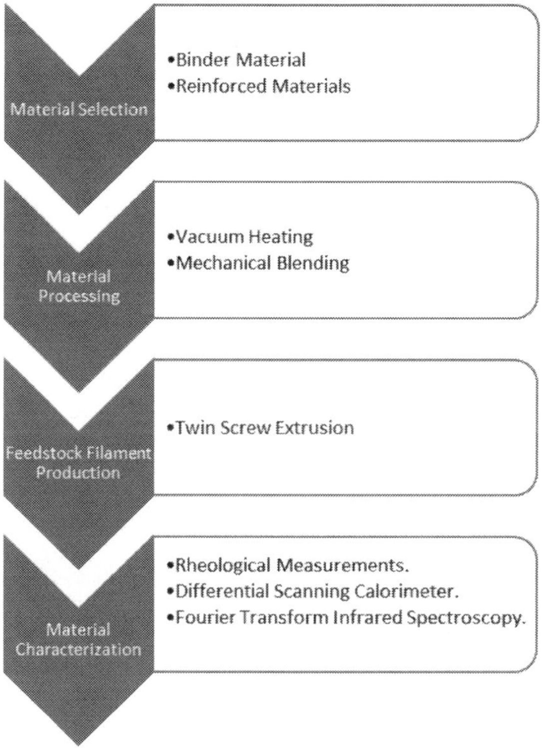

Figure 5.7 Process flow diagram.

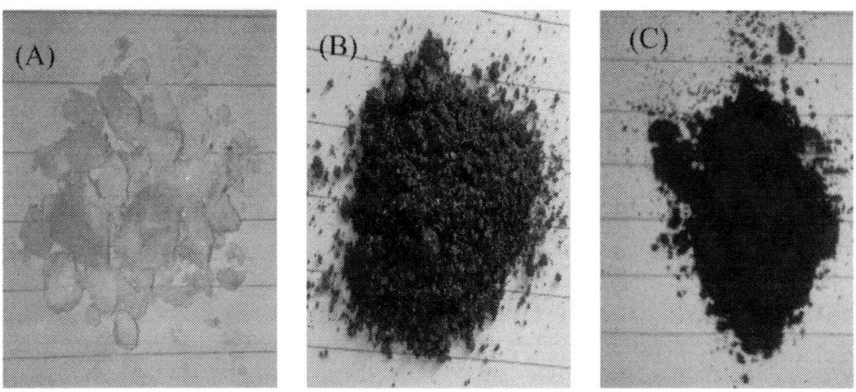

Figure 5.8 Feedstock filament materials (A) paraffin wax, (B) graphite, and (C) graphene.

Table 5.2 Weight proportion of the compositions.

Composition	Paraffin wax (by weight %)	Graphene (by weight %)	Graphite (by weight %)
A	50	50	–
B	60	40	–
C	70	30	–
D	50	–	50
E	60	–	40
F	70	–	30

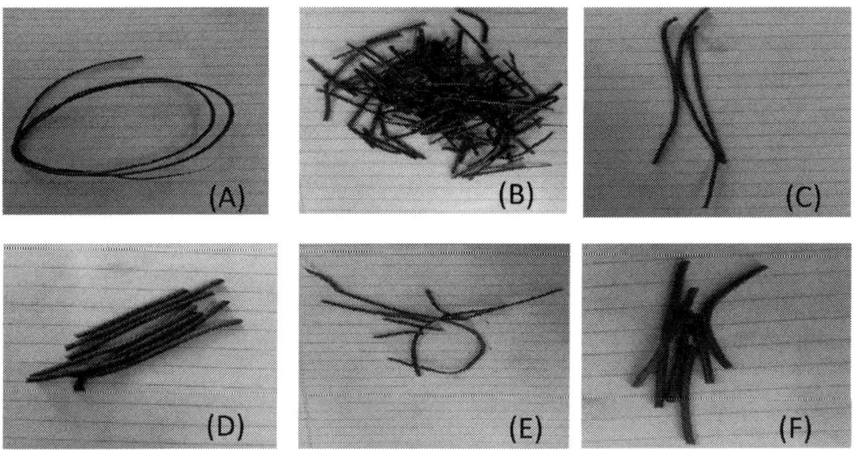

Figure 5.9 Extruded feedstock filament prepared with (A) composition A, (B) composition B, (C) composition C, (D) composition D, (E) composition E, and (F) composition F (as per Table 5.2).

5.2.2 Sample preparation

In order to remove the moisture, the selected materials were heated in vacuum oven (at a temperature 40°C) for 2 hours. Table 5.2 shows the compositions and proportion of paraffin wax, graphite powder, and graphene powder. The compositions were prepared by mechanical blending.

5.2.3 Sample processing

TSE has been used for the fabrication of feedstock filament (Fig. 5.9A–F). Initially, mechanical blending was carried out for the uniform mixing of composition through TSE. The processing parameters such as screw rotation (100 RPM) and the barrel temperature (40°C–50°C) were specified after carrying out initial experimental trials.

5.2.4 Materials characterization

5.2.4.1 *Rheological measurements*

MFI test has been conducted to investigate rheological characteristics of the prepared samples. MFI is designated as a standard technique to scrutinize the rheological behavior for comparative purpose. In this method, flow of thermoplastics materials passes through an orifice for 10 minutes are measured. The test was conducted at a 40°C with applied weight (1 kg) as per ASTM D1238 method (Fig. 5.10). Standard test units for MFI are g/10 minutes. Prepared samples are extruded through the barrel of MFI and weighted 3 times and average value is recorded.

MFI values and viscosity (calculated as per the procedure given by Shenoy et al. [15]) of prepared samples are tabulated in Table 5.3 and illustrated in Fig. 5.11.

5.2.4.2 *Differential scanning calorimetry*

DSC is a thermal analysis method in which the amount of heat required to increase the temperature of a sample and reference is measured as a function of temperature. Initially, the testing starts by maintaining the equal temperature of both the sample and reference, throughout the

Figure 5.10 Melt flow tester.

Table 5.3 MFI and viscosity values for different compositions/proportions (by weight %).

Composition	Density (ρ) g/cm³	MFI g/(10 min)	Viscosity (μ) in (Pa-s)
A	0.25	1.618	771
B	0.30	1.536	974
C	0.36	1.428	1257
D	1.28	1.327	4812
E	1.18	1.466	4015
F	1.09	1.517	3584

Note: The average value of three repeated observations has been quoted.

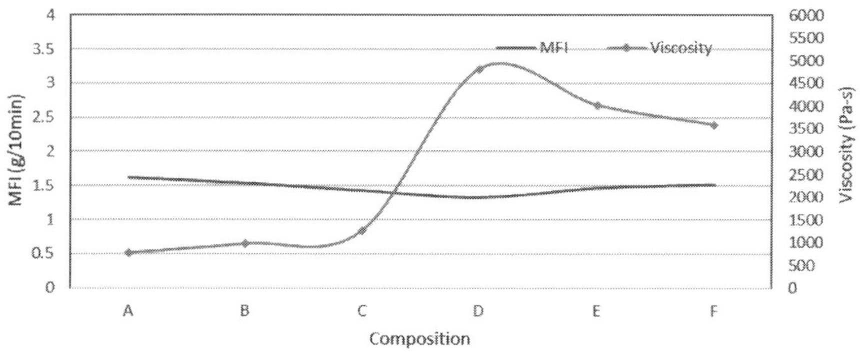

Figure 5.11 MFI versus viscosity graph of compositions (A–F).

experiment. It is used to determine fusion and crystallization as well as glass transition temperatures. The thermal properties of the prepared sample were recorded using DSC1 (Mettler Toledo) with HSS8 sensors and Huber TC intercooler (with an attachment of refrigerated cooling). Before starting, the samples are weighted about 6.8–8.2 mg in 40 μL Al standard crucible. The samples started heating from 30°C to 80°C with 1 min isothermal curing at 80°C and cooled to 30°C at a rate of 10 K/min. A second heating scan was also performed at the same rate (10 K/min). The measurement curves indicate the peak, whose area corresponds to the enthalpy involved in the process. The DSC curves include exothermic and endothermic peaks, namely, glass transition temperature (T_g), peak due to cold crystallization, melting enthalpy, and finally decomposition. The crystallinity value for both on melting (X_m) and cooling (X_c) was determined by using following equations [14]. The melting enthalpy (ΔH) of 100% crystalline paraffin wax is taken as 205.

Table 5.4 DSC results.

Material/composition	T_c (°C)	T_m (°C)	$\Delta H°_c$ (J/g)	$\Delta H°_m$ (J/g)	X_c (%)	X_m (%)
Paraffin wax	58.01	49.00	24.37	2.64	0.23	0.260
A	52.34	53.01	14.85	0.64	0.14	0.062
B	50.09	52.84	14.84	0.49	0.12	0.039
C	49.97	55.56	8.74	4.76	0.60	0.330
D	51.19	52.52	-5.67	0.32	0.50	0.031
E	54.23	53.70	4.87	0.48	0.03	0.039
F	55.06	56.37	8.7	2.66	0.60	0.180

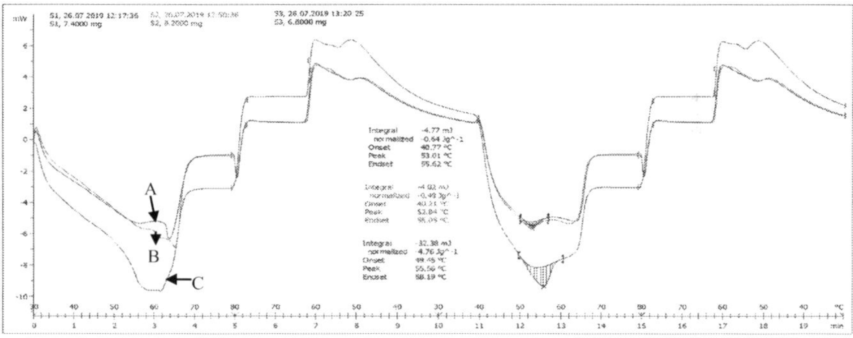

Figure 5.12 DSC thermographs compositions (A–C).

Table 5.4 shows the DSC results which include melting peak temperature (T_m), crystallization peak temperature (T_c), melting enthalpy ($\Delta H°_m$), crystallization enthalpy ($\Delta H°_c$), degree of crystallinity on heating (X_m), and degree of crystallinity on cooling (X_c) [14]. Thermographs of composition A–C (Fig. 5.12) and composition D–F (Fig. 5.13) highlight different peaks.

5.2.4.3 Fourier-transform infrared spectroscopy

FT-IR (Perkin Elmer make), spectrum two, having a wavelength (400–4000 cm^{-1}) and has been carried out at a room temperature (25°C). FT-IR is an analytical measurement method, used for the analysis of chemical structure. The IR radiation is passed through a chemical composition, some radiations are absorbed by the sample and some transmitted through it. The FT-IR graphs for the compositions A, B, and C are shown in Fig. 5.14. Similarly, for the compositions D, E, and F are shown in Fig. 5.15.

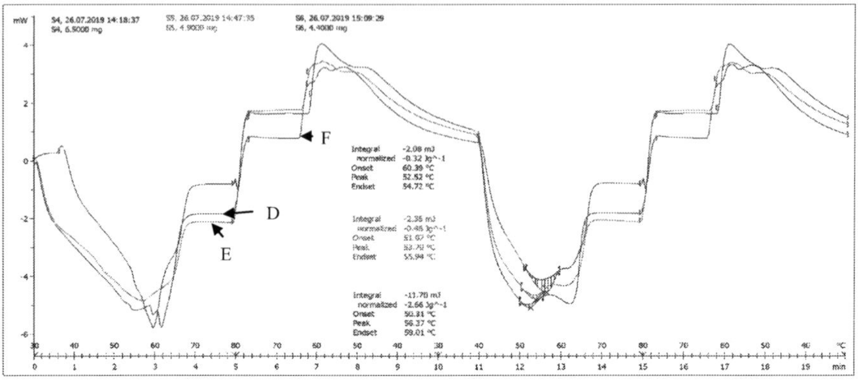

Figure 5.13 DSC thermographs of compositions (D–F).

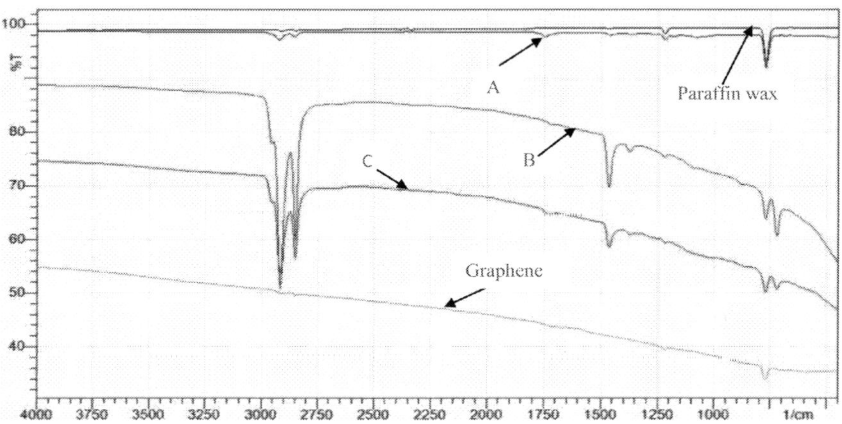

Figure 5.14 FT-IR spectra of composition A–C, paraffin wax and graphene.

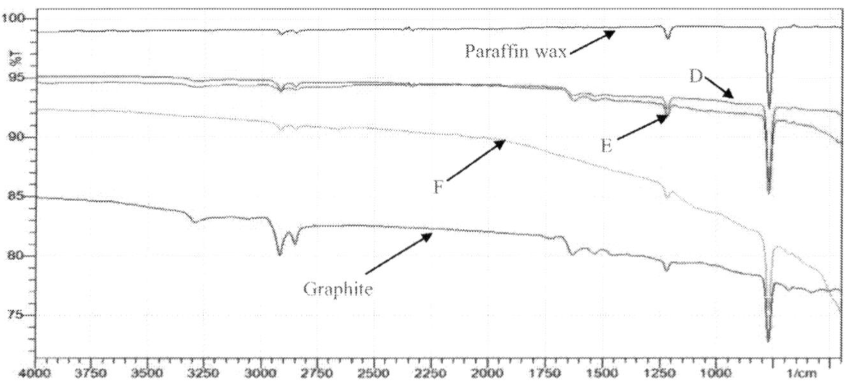

Figure 5.15 FT-IR Spectra of composition D–F, paraffin wax and graphite.

Table 5.5 FT-IR spectral data of composition A–C, paraffin wax, and grapheme.

A	B	C	Graphene	Paraffin wax	Assignment/compound class
2921.44	2954.37	2955.03	2913.58	2915.61	O-H stretching mode
2852.35	2915.91	2915.93	2845.58	2847.87	O-H stretching mode
–	2848.31	2848.33	–	–	O-H stretching mode
–	–	2660.78	–	–	O-H stretching mode
2376.63	–	–	–	2362.63	C≡N stretching mode/nitrile
–	–	–	–	2339.54	C≡N stretching mode/nitrile
1744.78	–	–	–	–	C = O stretching mode (Ester)
–	1723.12	–	–	–	C = O stretching mode (conjugation of Ester with C = C or phenyl)
1457.43	1463.10	1463.98	–	–	N-H: Bending in secondary amine
–	1374.35	1375.73	–	–	C-O-H: bending mode
1218.85	1217.68	1218.65	1218.25	1219.39	C-O stretching mode
1180.53	–	–	–	–	C-O stretching mode
1081.89	–	–	–	–	C-F alkyl halide
772.14	771.84	771.76	771.88	772.16	C-H (out-of-plane deformation vibration)
–	720.44	721.70	–	–	= C-H: out-of-plane bending mode (Aromatic ring)
–	–	–	684.21	–	C-CI: stretching mode (alkyl halide)
–	–	–	592.50	–	C-Br: stretching mode (alkyl halide)
–	–	–	512.07	–	C-Br: stretching mode (alkyl halide)

As shown in Table 5.5, the graphene showed the peak at the range 512.07 and 592.50, due to the presence of compound class alkyl halide and C-Br, having a stretching mode characteristic. The graphene peak also indicated at 684.21, which shows characteristics C-CI: stretching mode with Alkyl Halide compound class. In case of composition "B" and "C," the peaks at 720.44 and 721.70, respectively, shows = C-H, aromatic ring characteristics (out-of-plane bending mode) with Alkene as compound class which is absent from other compositions as well as in graphene and paraffin wax. Moreover, compositions "A," "B," "C," graphene and paraffin wax show peak at a value of 772.14, 771.84, 771.76, 771.88, and 772.16, respectively, that indicates C-H, out-of-plane deformation vibration characteristics with alkene compound class. The peak at 1081.89 in the case of composition "A" is associated with characteristics band C-F and compound class alkyl halide. The peak value 1180.53 in material is related to C-O stretching mode characteristics band and alcohol compound class. Composition "A," "B," "C," graphene and paraffin wax at peak value 1218.85, 1217.68, 1218.65, 1218.25, and 1219.39, respectively, represents C-O stretching mode characteristics band and compound class alcohol. The peak at value 1374.35, 1375.73 in composition "B" and "C," respectively, shows association with C-O-H: bending mode characteristics band which are not visible in composition "A," graphene and paraffin wax. The peak value at 1457.43, 1463.10, and 1463.98 represents N-H: Bending in secondary amine in material "A," "B," and "C."

The peak value 1723.12 in composition "B" showed convective bonding due to C = O stretching mode characteristics with conjugation of ester with C = C or phenyl. The peak at 1744.78 in composition "A" showed the C = O stretching mode characteristics band and ester compound class. The peak value 2376.63 and peak range 2339.54–2362.63 in composition "A" and "paraffin wax," respectively, represents C≡N stretching mode characteristics band which are not visualize in material "B," "C," and graphene. The peak range 2921.44–2852. 35 in composition "A," the peak range 2954.37–2848.31 in composition "B," the peak range 2955.03–2660.78 in material "C," The peak range 2913.58 2845.58 in graphene, and the peak range 2915.61–2847.87 in paraffin wax represent relation with compound class carboxylic acid and presence of O-H stretching characteristics band. After analyzing FT-IR spectral data, it validates the successful chemical structure integration and reinforcement of graphene in paraffin wax in varying proportions.

The FT-IR spectral data of composition D-F, paraffin wax, and graphite have been tabulated in Table 5.6. The composition "D" shows peak at the range 504.32–592.50, represents the presence of C-Br, stretching mode characteristics and compound class is alkyl halide. At peak values of 682.65, 680.62, and 683.07 in case of compositions "D," "E," and "F," respectively, shows characteristics C-Cl: stretching mode with alkyl halide compound class. The peak value at 772.05, 772.13, 772.15, 772.07, and 772.16 in composition "D," "E," "F," graphite, and paraffin wax respectively is associated with C – H, out-of-plane deformation vibration. The peak value at 1219.20, 1219.17, 1219.36, 1218.40, and 1219.39 in composition "D," "E," "F," graphite, and paraffin wax is associated with compound class ester and characteristics band C = O stretching mode. The peak value 1455.92 in

Table 5.6 FT-IR spectral data of composition D-F, paraffin wax, and graphite.

D	E	F	Graphite	Paraffin wax	Assignment/compound class
3292.05	–	–	–	–	O-H stretching mode
–	3286.28	3286.06	2916.70	2915.61	O-H stretching mode
3061.67	2917.50	2916.96	2848.80	2847.87	O-H stretching mode
2918.22	2849.93	–	2661.08	–	O-H stretching mode
2850.32	–	–	–	–	O-H stretching mode
–	2363.42	2362.77	–	2362.63	C≡N stretching mode/nitrile
–	–	2338.65	–	2339.54	C≡N stretching mode/nitrile
–	–	–	2168.22	–	C≡C: stretching mode (Alkyne)
1730.55	–	–	–	–	C = O stretching mode (conjugation of ester with C = C or phenyl)
1632.18	1625.81	1624.44	–	–	N-H: bending in primary amine
1536.36	1537.86	1538.82	–	–	-NO2 asymmetric stretching mode (aliphatic nitro)
1455.92	–	–	–	–	N-H: bending in secondary amine
1219.20	1219.17	1219.36	1218.40	1219.39	C = O stretching mode (Ester)
772.05	772.13	772.15	772.07	772.16	C - H out-of-plane deformation vibration
682.65	680.62	683.07	–	–	C-CI: stretching mode (alkyl halide)
583.15	–	–	–	–	C-Br: stretching mode (alkyl halide)
504.32	–	–	–	–	C-Br: stretching mode (alkyl halide)

composition "D" is related to N-H, which is a bending characteristic in secondary amine. The peak value 1730.55 in composition "D" showed convective bonding due to C = O stretching mode characteristics due to conjugation of ester with C = C or phenyl. The peak value at 2168.22 in graphite is related to C≡C that is stretching mode characteristics band and compound class alkyne. The peak value 2363.42, the peak range 2362.77–2338.65 and the peak range 2362.63–2339.54 in composition "E," "F," and paraffin wax, respectively, associated with C≡N stretching mode characteristics band which is not visible in composition "D" and Graphite. Further, the peak range 3061.67–2850.32, 3286.28–2849.93, 3286.06–2916.96, 2916.70–2661.08, and 2915.61–2847.87 in composition "D," "E," "F," graphite, and paraffin wax, respectively, shows relations with

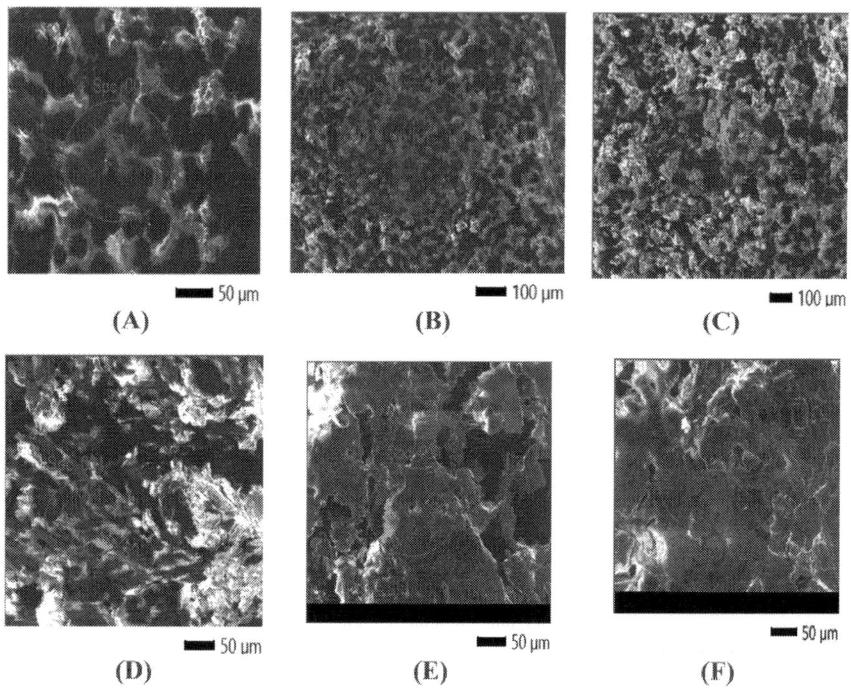

Figure 5.16 SEM images, (A) composition A, (B) composition B, (C) composition C, (D) composition D, (E) composition E, and (F) composition F.

compound class carboxylic acid and characteristics band O-H stretching mode. In composition "D," peak value 3292.05 showed the O-H stretching mode characteristics band and Alcohol compound class which is absent from other compositions, graphite and paraffin wax. There is some slight variation between the different proportions, these differences are possibly due to the varying compositions. After inspecting FT-IR spectral data, it is confirmed graphite is reinforced in paraffin wax and chemical structure changes successfully.

5.2.4.4 Scanning electron microscopy

Structural morphology was assessed by using SEM images of prepared sample as illustrated in Fig. 5.16A–F, respectively.

5.3 Conclusion

Alternative feedstock filament materials for FDM have been successfully proposed for the energy storage applications. The characterization of reinforced nanomaterial in polymeric material matrix (graphene and graphite in paraffin wax) has been carried out by MFI and DSC. Further, the structural changes have been observed by FT-IR and SEM images. The following observations are noted from this experimental research work;

1. The web of science-based research gap analysis has been carried out to identify the possible scope of the research work. Limited work has been carried out for the 3DP of energy storage devices.
2. The MFI value and viscosity of different proportions of graphene and graphite in paraffin (composition A-F) have been plotted, in order to understand the flow behavior and establishing the processing conditions such as liquefier head temperature and material flow rate for the 3DP of direct energy storage devices.
3. The DSC results reveals that, the values of T_C, T_m, X_m, and X_c have been decreased with the inclusion of reinforced materials in paraffin wax matrix. Although the nucleation effect also observed in the matrix but the reinforcement slightly resists the particle movement and diffusion of paraffin wax molecular chains to the surface of the nucleus in the compositions.
4. Fourier-transform infrared spectroscopy study illustrates the changes observed in the chemical structure with the inclusion of graphene and graphite in paraffin wax. Further, FT-IR data confirm the reinforcements in paraffin wax.
5. Finally, the SEM images show the uniform distribution of different reinforced materials in the mixture. Moreover, no lumps or agglomeration are available in the matrix. Hence, the proposed composite may be used for 4D applications.
6. Further research work may be focused on the printing of functional parts of energy storage device such as dry cell, having tailor-made geometry and capacity.

Acknowledgment

The authors are thankful to Institution of Engineers (India) for financial support under project ID: UG2019014.

References

[1] Li H, Peng L, Zhu Y, Zhang X, Yu G. Achieving high-energy—high-power density in a flexible quasi-solid-state sodium ion capacitor. Nano Letters 2016;16(9):5938—43.

[2] González E, Goikolea JA, Barrena R, Mysyk. Review on supercapacitors: technologies and materials. Renewable & Sustainable Energy Reviews 2016;58:1189–206.
[3] Boparai KS, Singh R. 3D printed functional prototypes for electrochemical energy storage. International Journal of Materials Engineering Innovation 2019;10(2):152–64.
[4] Sun Y, Lopez J, Lee H-W, Liu N, Zheng G, Wu C-L, et al. A stretchable graphitic carbon/si anode enabled by conformal coating of a self-healing elastic polymer. Advanced Materials 2016;28:2455–61.
[5] Lee C-Y, Taylor AC, Nattestad A, Beirne S, Wallace GG. 3D print electrocatalytic applications. Joule 2019;3(8):1835–49.
[6] Strickler AL, Higgins D, Jaramillo TF. Crystalline strontium iridate particle catalysts for enhanced oxygen evolution in acid. ACS. Applied Energy Materials 2019;2(8):5490–8.
[7] Zhang CFJ, Kremer MP, Seral-Ascaso A, Park S-H, McEvoy N, Anasori B, et al. Stamping of flexible, coplanar micro-supercapacitors using MXene Inks. Advanced Functional Materials 2018;28:1705506.
[8] Foster C, Down M, Zhang Y, et al. 3D printed graphene based energy storage devices. Scientific Reports 2017;7:42233.
[9] Godwin I, Rovetta A, Lyons M, Coleman J. Electrochemical water oxidation: the next five years. Current Opinion in Electrochemistry 2018;7:31–5.
[10] King LA, Hubert MA, Capuano C, et al. A non-precious metal hydrogen catalyst in a commercial polymer electrolyte membrane electrolyser. Nature Nanotechnology 2019;14:1071–4.
[11] Osiak M, Geaney H, Armstrong E, O'Dwyer C. Structuring materials for lithium-ion batteries: advancements in nanomaterial structure, composition, and defined assembly on cell performance. Journal of Materials Chemistry A 2014;2:9433–60.
[12] Rolison DR, Long JW, Lytle JC, Fischer AE, Rhodes CP, McEvoy TM, et al. Multifunctional 3D nanoarchitectures for energy storage and conversion. Chemical Society Reviews 2009;38:226–52.
[13] Boparai KS, Singh R, Singh H. Modeling and optimization of extrusion process parameters for the development of Nylon6–Al–Al 2 O 3 alternative FDM filament. Progress in Additive Manufacturing 2016;1(1–2):115–28.
[14] Boparai KS, Singh R, Fabbrocino F, Fraternali F. Thermal characterization of recycled polymer for additive manufacturing applications. Composites Part B: Engineering 2016;106:42–7.
[15] Shenoy AV, Saini DR, Nadkarni VM. Melt rheology of polymer blends from melt flow index. International Journal of Polymeric Materials 1984;10(3):213–35.

CHAPTER 6

On dual/multimaterial composite matrix for smart structures: a case study of ABS-PLA, HIPS-PLA-ABS

Rupinder Singh[1], Sudhir Kumar[2] and Ranvijay Kumar[3]
[1]Department of Mechanical Engineering, National Institute of Technical Teachers Training and Research, Chandigarh, India
[2]Department of Mechanical Engineering, CT University, Ludhiana, India
[3]Department of Mechanical Engineering, University Center for Research and Development, Chandigarh University, Mohali, India

6.1 Introduction

The fused deposition modeling (FDM) is one of the low-cost additive manufacturing processes that enables the manufacturer to customize the design as well and materials. Apart from the cost of the final product manufactured by the FDM process, the manufactured product is highly criticized due to lower flexibility, functionalities, and surface properties. The product manufactured by using single materials may have limited features and functionalities. But nowadays, multimaterial 3D printing using FDM technology has overcome with the restriction of functions and features. The advancement in multimaterial 3D printing has opened new innovative applications for robotics, bio-manufacturing, actuators, sensors, space, and various engineering applications. The previous studies have suggested that the strain recovery product can be prepared by using multimaterial 3D printing process using FDM process [1]. Li et al. [2] have used FDM 3D printing for single material, multimaterial, and print-pause-printing concept. The study suggested that multimaterial 3D printing process has been used to increase the functionalities of microchemical devices [2]. Similarly, the studies have reported the applications of multimaterials 3D printing for preparation of machine vision assisted platform [3], microfluidic [4], actuators [5], 4D printing [6], soft pressure sensors [7], smart devices [8], and biomedical [9] applications. The remarkable advancements have been reported by multimaterial 3D printing process which is

4D Printing
DOI: https://doi.org/10.1016/B978-0-12-823725-0.00002-3

not possible by single materials 3D printing process. For example, the researchers have simulated and designed the multistate structure monolithically using multimaterial 3D printing [10]. The multimaterial 3D printing process has enabled the preparation of bidirectional and stretchable piezoresistive sensors of thermoplastic polyurethane—carbon nanotube combination [11]. Also, the 3D printing has been reported for preoperative planning of surgery [12]. The multimaterial 3D printing is not only limited for the polymers, metals, or alloys but nowadays it has been largely used for the 3D printing of food with simultaneous infrared cooking [13]. The study has reported the 3D food printing by multimaterials 3D printing process with complex internal structure suitable for conventional postprocessing [14]. Also, the multimaterial 3D printing process continues for preparation of customized foods [15]. The previous studies have suggested that the multimaterial 3D printing process can helpful to modify the tensile, flexural, pullout, and surface properties [16—18]. In the field of sensor making, the multimaterial 3D printing has played an important role. The study has been reported on preparation of stretchable tactile sensors [19], resistive sensors for soft robotics [20], instrumented cardiac micro-physiological devices, optoelectronics [21], and compressive sensing [22].

The detailed bibliographic studies have been conducted by using the web of science database between years of 2016 and 2020. A total of 259 research papers have been reported by putting the keyword "multimaterial 3D printing." The binary counting algorithm has been selected using VOS Viewer software package, and a total of 7603 terms have been found and out of those 96 have met the threshold. About 60% terms have been selected, that is, 58 on set algorithm and among those 26 have been left after refinement. The relevance score and occurrences of terms have been presented in Table 6.1. Fig. 6.1 shows the bibliographic map analysis of the selected terms. The "manufacturing" term has been found mostly in all reported studies as per the visual node size. It has been found that actuators keywords reported many times. This means that the multimaterial 3D printing has wide scope in 4D printing of sensors and actuators. The gap in the studies can be predicted from Fig. 6.2. As observed from Fig. 6.2, multimaterial 3D printing has been less explored in actuator and sensors applications. The studies may be conducted to 4D printing of multimaterial components for actuator and development of sensors.

Fig. 6.3 show the methodology used for dual-material printing and derive thumb rule for multimaterial combination selection criterion.

Table 6.1 Selected term of their occurrences and relevance score.

Number	Term	Occurrences	Relevance score
1	Accuracy	21	0.798
2	Actuator	22	1.0331
3	Complexity	22	0.9122
4	Composite	25	0.5524
5	Direct ink writing	14	1.0088
6	Electronic	25	1.0899
7	Extrusion	16	1.1483
8	Flexibility	19	0.5519
9	Function	31	0.2565
10	Functionality	30	0.7611
11	Hydrogel	26	1.3412
12	Integration	24	1.0065
13	Manufacturing	77	0.4128
14	Mechanical property	26	1.0061
15	Multimaterial printing	11	1.0607
16	Multimaterial structure	13	0.9089
17	Multiple material	13	1.7085
18	Organ	14	1.6901
19	Performance	36	0.4995
20	Self	14	0.8
21	Sensor	20	1.4582
22	Soft robotic	19	1.2353
23	stereo lithography	18	0.411
24	Strength	19	1.1384
25	Temperature	21	0.8385
26	Tissue engineering	15	2.372

6.2 On dual-material 3D printing of different combination of layers: a case study

A case study was previously conducted on dual-material printing of acrylonitrile butadiene styrene (ABS) and polylactic acid (PLA) material on 3D printing platform of FDM [23]. In that study 12 different combinations of ABS and PLA material deposition were used (Table 6.2) for carrying out analysis especially for checking the effect for dual-material combination on mechanical properties of 3D printed specimen. Fig. 6.4 shows the 3D printed samples of ABS and PLA-based dual-material specimen. The 3D printed samples were tested for mechanical properties using UTM testing machine (Make: Shanta Engineering, Pune, India) and Fig. 6.5A–D shows the observed mechanical properties for different 3D

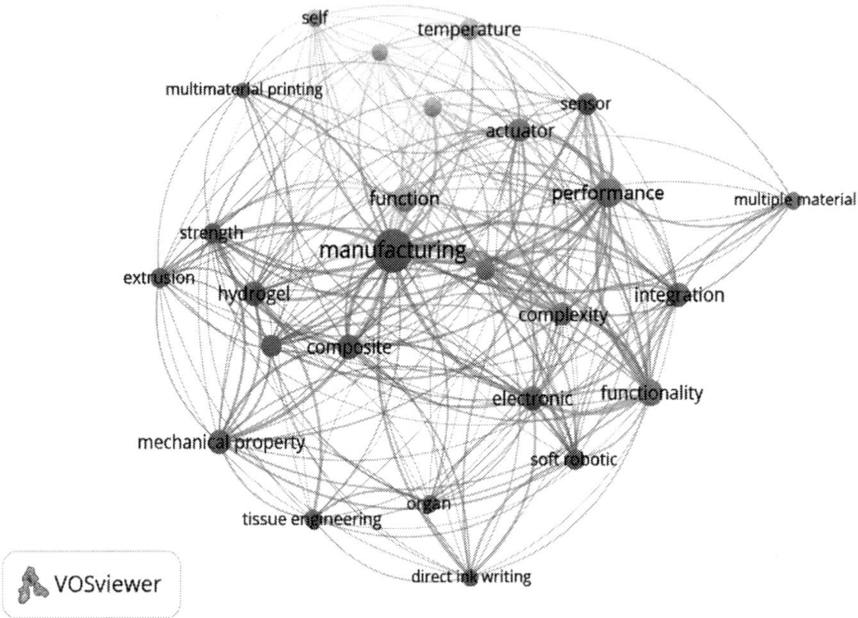

Figure 6.1 Detailed bibliographic map for multimaterial 3D printing.

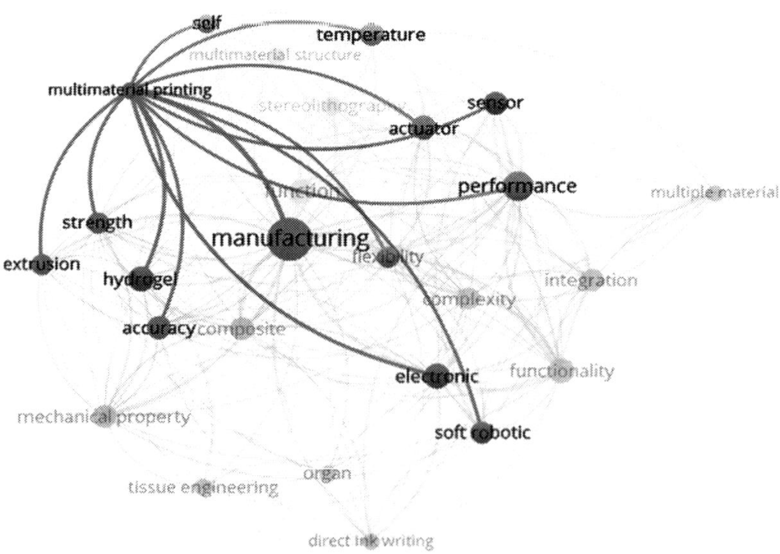

Figure 6.2 Bibliographic gap analyses for multimaterial 3D printing.

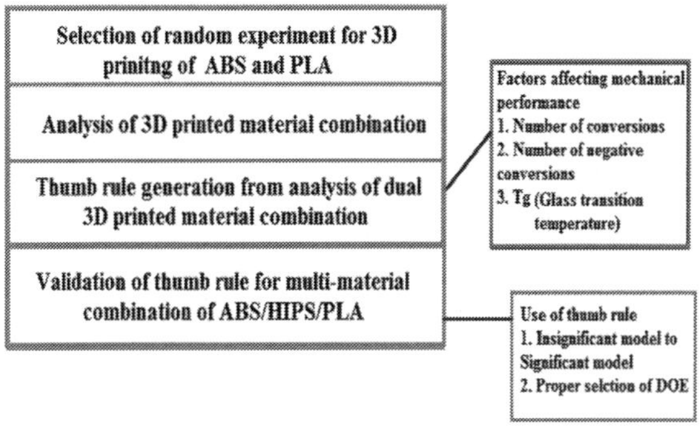

Figure 6.3 Methodology used for present case study.

Table 6.2 Different combination of ABS and PLA for 3D printing of tensile specimen.

S. No.	Combination of layers
1	PPPPPP
2	AAAAAA
3	PAPPAP
4	PPPAAP
5	PPPPAA
6	APAAPA
7	AAAPPA
8	AAAAPP
9	APAPAP
10	PPAPAA
11	AAPAPP
12	PPPAAA

Note: P: PLA and A: ABS.

printed sample. From UTM testing, it has been observed that the maximum strength was observed for single material-based sample of PLA (Fig. 6.5C) while for other specimens the combination of layer has played significant role in determining the mechanical performance of 3D printed specimen. It was observed that there were two different types of deposition for the selected different combination of 3D printing (1) deposition of PLA on ABS and (2) deposition of ABS on PLA layer. The effect of two different types of deposition was analyzed, and it was observed that the second type of deposition in which ABS was deposited on PLA has shown better result than its counterpart. Thus the deposition of PLA on ABS has been observed as negative conversion (NC) while printing of

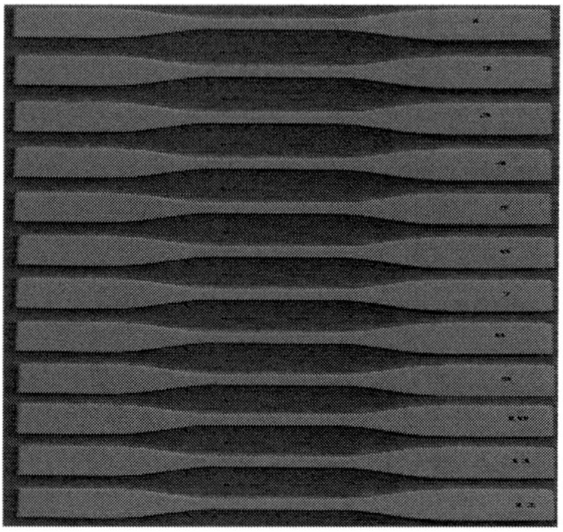

Figure 6.4 3D printed tensile specimen for dual-material combination of ABS and PLA.

ABS on PLA has been observed as positive conversion (PC). Thus in as sample where different number of layers to be selected for 3D printing of specimens the effect of NC and PC must be taken care of otherwise the final mechanical performance of 3D printed sample may be poor. This may be attributed to the fact that ABS has higher glass transition temperature (T_g) (109.65°C) than PLA (64°C). Thus when ABS was printed on PLA the side flow existed and thus resulted into proper fusion of material with PLA while when PLA was printed on ABS the low T_g of PLA significantly reduces the layer temperature within short interval of time thus resulting into poor fusion of PLA on ABS substrate.

When there is more than one conversion or multimaterial specimen needs to be printed, then one must care of number of negative conversions (NoNC) and number of conversions (NoC). Based on observed behavior and existence of NOC and NoNC, Table 6.3 shows the NONC and NoC of the selected combination of dual-material printed ABS and PLA.

6.3 Thumb rule derived from the case study

From the observed trends for dual material-based 3D printing, the thumb rule was derived for 3D printing of multimaterial specimens where one must consider T_g of material while choosing the combination for 3D printing. Three important parameters (Tg, NoC, and NoNC) which

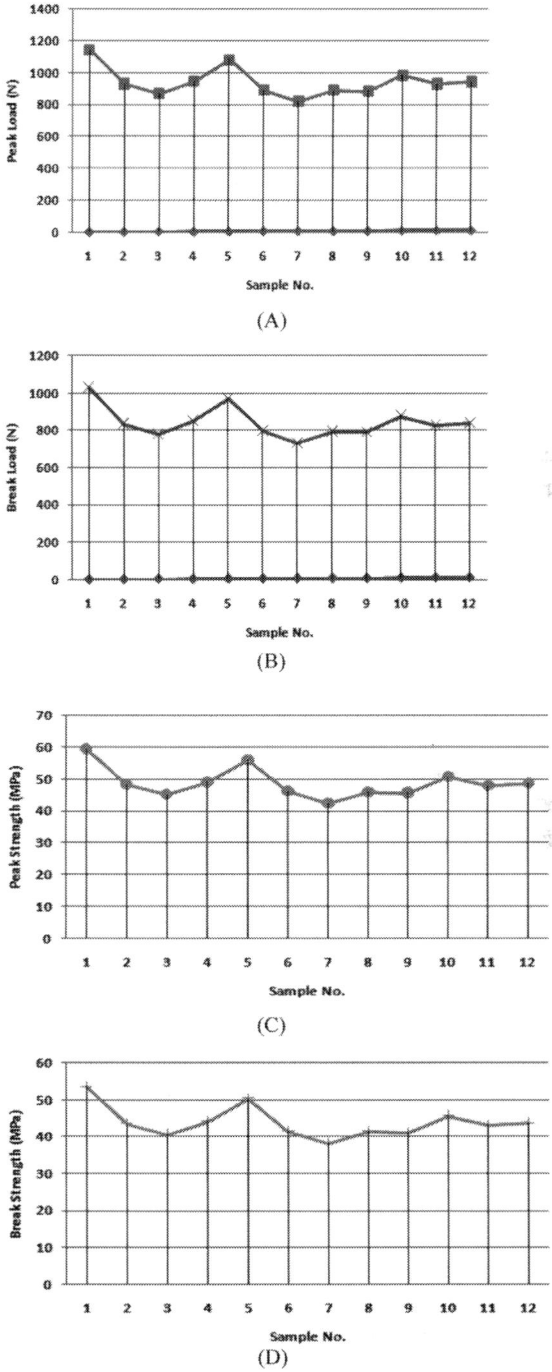

Figure 6.5 UTM tested dual material-based 3D printed samples for (A) peak load, (B) break load, (C) peak strength, and (D) break strength.

Table 6.3 NoC and NoNC for selected combination of BAS and PLA.

S. No.	Combination of layers	NoC	NoNC
1	PPPPPP	0	0
2	AAAAAA	0	0
3	PAPPAP	4	2
4	PPPAAP	2	1
5	PPPPAA	1	0
6	APAAPA	4	2
7	AAAPPA	2	1
8	AAAAPP	1	1
9	APAPAP	5	3
10	PPAPAA	3	1
11	AAPAPP	3	2
12	PPPAAA	1	0

played significant role in mechanical performance of specimen; therefore, proper care must be given while selecting the design of experimentation (DOE). One must consider that NoNC will always lead to poor performance, thus NoNC must be avoided while selecting the DOE for any multimaterial-based 3D printing study.

6.4 Validation of thumb rule

6.4.1 Case study for validation of thumb rule from dual-material 3D printing on 3D printed three different material combinations of ABS/PLA and high-impact polystyrene

Kumar et al. in [18,23] performed 3D printing of different combination of ABS, PLA, and high-impact polystyrene (HIPS) material for structural engineering applications. Table 6.4 shows the results observed by Kumar et al. [18,23] for different combination of ABS/PLA/HIPS used for 3D printing with chosen DOE for 3D printing. It was observed from this study that the infill speed has played maximum contribution toward the output while second rank was given to infill density and third rank was provided to material combination, and moreover, no parameter from this study was significant as their p value was higher than 0.05. Thus it may be concluded that the study was performed without proper selection of DOE where no consideration was given to selection of 3D printing of material combination and randomly material was 3D printed on each other resulting into insignificant process parametric study.

Table 6.4 DOE used for 3D printing of ABS/PLA/HIPS.

Selected DOE				Output obtained	
S. No.	Layer combination	Infill density	Infill speed	Break load (N)	Break strength (MPa)
1	APH	60	50	120	6.28
2	APH	80	60	161	8.43
3	APH	100	70	186	9.70
4	PHA	60	60	145	7.56
5	PHA	80	70	170	8.90
6	PHA	100	50	169	8.81
7	HAP	60	70	134	6.98
8	HAP	80	50	157	8.19
9	HAP	100	60	147	7.71

Table 6.5 NoNC and NoC of selected material combination in Table 6.4.

S. No.	Layer combination	NoC	NoNC
1	APH	2	1
2	APH	2	1
3	APH	2	1
4	PHA	2	0
5	PHA	2	0
6	PHA	2	0
7	HAP	2	1
8	HAP	2	1
9	HAP	2	1

Note: A: ABS; P: PLA and H: HIPS.

It has been observed that if the proper criterion (based on NoNC, Tg, and NoC) for 3D printing of multimaterial was to be taken care of only then the study may result into significant results with better mechanical performance in comparison with reported observations by Kumar et al. [18,23]. In this case, NoNC was 3D printing of PLA on ABS or on HIPS for three material combinations (as T_g (ABS) > T_g (HIPS) > T_g (PLA)). From Table 6.5, it was observed that the NoNC in case of samples 4–6 was zero thus expected to show best mechanical performance among all but the observed trends (Table 6.4) has shown that the sample 3 has shown best results as it was printed with 100% infill density. It may be observed that the samples 5 and 6 have been printed with 80 and 100% infill density, respectively, and have shown good mechanical performance.

But sample 6 in which there was zero NoNC and 100% infill density has shown third rank among all 3D printed sample as it was printed with 50 mm/s infill speed. Thus we may conclude that the DOE lacks in selection of proper material and input parameter combination thus has shown insignificant results. Thus if one needs to make the model significant proper selection of DOE was important before carrying out the experimentation. As infill speed and infill density has important role to play in output properties one must select proper DOE considering NoNC and NOC with other input parameters for 3D printing of multimaterial specimen. The 3D printed samples were examined under tool maker microscope, and it was observed that when PLA was printed on ABS or HIPS (considered as NoNC), there existed a side flow of material which resulted into poor fusion of material and ultimately poor mechanical performance (Fig. 6.6). The 3D rendered image and surface roughness profile for sample 1 and 4 confirms the observation that 3D printing of material with NoNC (sample 1) results into poor surface roughness as its Ra value was 59.36 nm, whereas for sample without any NoNC (sample 4), the surface roughness was less about 24.56 nm.

6.4.2 Proposed best and worst condition while considering NoC, NoNC, and other input parameters

1. While for best mechanical performance the infill density of 100%, infill speed of 70 mm/s and material combination of PHA may act as best 3D printed tensile specimen if selected in DOE.
2. While considering the NoC, NoNC, infill density, and infill speed, the proposed worst condition for three material-based combination of ABS, PLA, and HIPS was AHP in which there were two NoNC (deposition of H on A and then P on H), infill density 60%, and Infill speed 50 mm/s.

While the proposed best and worst conditions were outside the selected DOE (Table 6.3) of Kumar et al. [18,23], the optimized model was therefore insignificant.

6.5 Summary

The present study has been performed on multimaterial combination for proper selection of DOE supported by previously reported work on dual material-based 3D printed prototypes in which it was observed that the material combination must be judicially selected for better mechanical

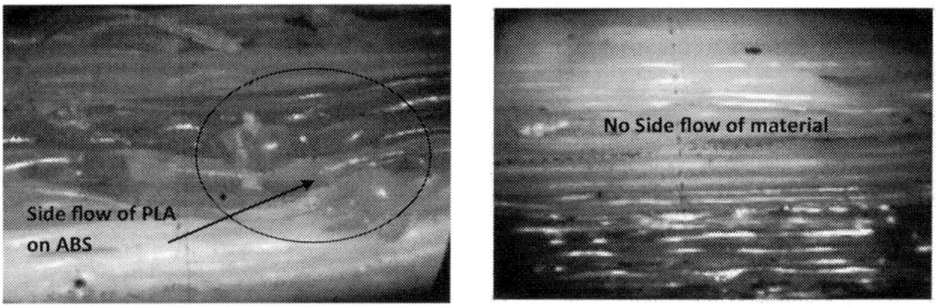

Sample no. 1(Base: ABS; Middle layer: PLA; Top Layer: HIPS)

Sample no. 4 (Base layer: PLA; Middle layer: HIPS; Top Layer: ABS)

(A)

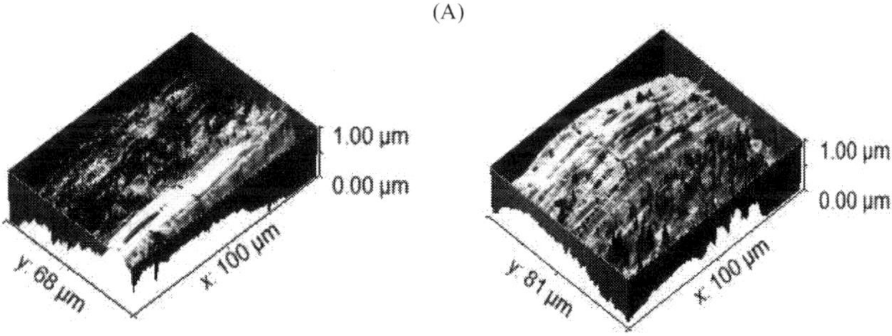

Sample no. 1

Sample no. 4

(B)

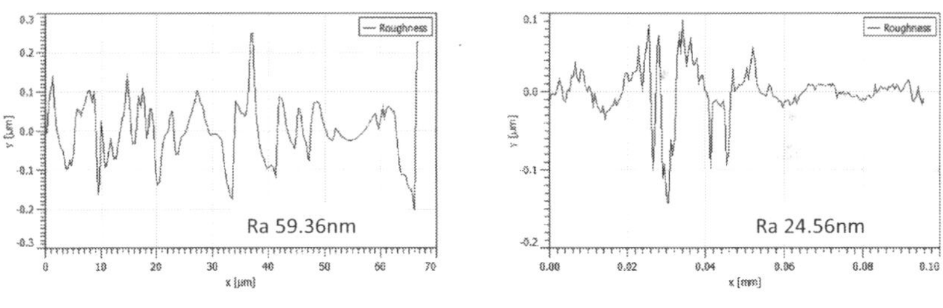

Sample 1

Sample no. 4

(C)

Figure 6.6 (A) Tool maker microscopic images for sample 1 and sample 4; (B) 3D rendered image for sample 1 and sample 4; and (C) surface roughness profile for samples 1 and 4.

performance. From the case study of dual material-based 3D printing, a thumb rule was derived which considers the NoC and NoNC based on T_g of material for better mechanical results. The observed thumb rule was validated on a previous study performed by Kumar et al. [18,23] in which their optimized model was insignificant and to make the model significant thumb rule was applied and best and worst conditions of 3D printing were proposed. Further the Tool maker microscopic analysis for 3D printed samples has also supported the derived thumb rule.

References

[1] Mansouri MR, Montazerian H, Schmauder S, Kadkhodapour J. 3D-printed multi-material composites tailored for compliancy and strain recovery. Composite Structures 2018;184:11–17.

[2] Li F, Macdonald NP, Guijt RM, Breadmore MC. Increasing the functionalities of 3D printed microchemical devices by single material, multimaterial, and print-pause-print 3D printing. Lab on a Chip 2019;19(1):35–49.

[3] Sitthi-Amorn P, Ramos JE, Wangy Y, Kwan J, Lan J, Wang W, et al. MultiFab: a machine vision assisted platform for multi-material 3D printing. ACM Transactions on Graphics (Tog) 2015;34(4):1.

[4] Begolo S, Zhukov DV, Selck DA, Li L, Ismagilov RF. The pumping lid: investigating multi-material 3D printing for equipment-free, programmable generation of positive and negative pressures for microfluidic applications. Lab on a Chip 2014;14 (24):4616–28.

[5] Zhang YF, Ng CJ, Chen Z, Zhang W, Panjwani S, Kowsari K, et al. Miniature pneumatic actuators for soft robots by high-resolution multimaterial 3D printing. Advanced Materials Technologies 2019;4(10):1900427.

[6] Tibbits S. 4D printing: multi-material shape change. Architectural Design 2014;84 (1):116–21.

[7] Emon MO, Alkadi F, Philip DG, Kim DH, Lee KC, Choi JW. Multi-material 3D printing of a soft pressure sensor. Additive Manufacturing 2019;28:629–38 Aug 1.

[8] Lee J, Kim HC, Choi JW, Lee IH. A review on 3D printed smart devices for 4D printing. International Journal of Precision Engineering Manufacturing-Green Technol 2017;4(3):373–83.

[9] Placone JK, Engler AJ. Recent Advances in Extrusion-Based 3D Printing for Biomedical Applications. Advanced Healthcare Materials 2018;7(8):1701161.

[10] Chen T, Mueller J, Shea K. Integrated design and simulation of tunable, multi-state structures fabricated monolithically with multi-material 3D printing. Scientific Reports 2017;7(1):1–8.

[11] Christ JF, Aliheidari N, Pötschke P, Ameli A. Bidirectional and stretchable piezoresistive sensors enabled by multimaterial 3D printing of carbon nanotube/thermoplastic polyurethane nanocomposites. Polymers 2019;11(1):11.

[12] Coelho G, Chaves TM, Goes AF, Del Massa EC, Moraes O, Yoshida M. Multimaterial 3D printing preoperative planning for frontoethmoidal meningoencephalocele surgery. Child's Nervous System 2018;34(4):749–56.

[13] Hertafeld E, Zhang C, Jin Z, Jakub A, Russell K, Lakehal Y, et al. Multi-material three-dimensional food printing with simultaneous infrared cooking. 3D Print Addititive Manufacturing 2019;6(1):13–19.

[14] Lipton J, Arnold D, Nigl F, Lopez N, Cohen D, Norén N, et al. Multi-material food printing with complex internal structure suitable for conventional post-processing. In Solid freeform fabrication symposium 2010;9:809–815.

[15] Sun J, Peng Z, Zhou W, Fuh JY, Hong GS, Chiu A. A review on 3D printing for customized food fabrication. Procedia Manufacturing 2015;1:308–19.

[16] Kumar S, Singh R, Singh M, Singh TP, Batish A. Multi material 3D printing of PLA-PA6/TiO2 polymeric matrix: flexural, wear and morphological properties. Journal of Thermoplastic Composite Materials 2020; Sep 2:0892705720953193.

[17] Kumar S, Singh R, Singh TP, Batish A. Flexural, pull-out, and fractured surface characterization for multi-material 3D printed functionally graded prototype. Journal of Composite Materials 2020;54(16):2087–99.
[18] Singh R, Kumar R, Farina I, Colangelo F, Feo L, Fraternali F. Multi-material additive manufacturing of sustainable innovative materials and structures. Polymers 2019;11(1):62.
[19] Guo SZ, Qiu K, Meng F, Park SH, McAlpine MC. 3D printed stretchable tactile sensors. Advanced Materials 2017;29(27):1701218.
[20] Shih B, Christianson C, Gillespie K, Lee S, Mayeda J, Huo Z, et al. Design considerations for 3D printed, soft, multimaterial resistive sensors for soft robotics. Frontiers in Robotics and AI 2019;6:30.
[21] Lind JU, Busbee TA, Valentine AD, Pasqualini FS, Yuan H, Yadid M, et al. Instrumented cardiac microphysiological devices via multimaterial three-dimensional printing. Nature Materials 2017;16(3):303–8.
[22] Tan JC, Low HY. Multi-materials fused filament printing with embedded highly conductive suspended structures for compressive sensing. Additive Manufacturing 2020;36:101551.
[23] Kumar R, Singh R, Farina I. On the 3D printing of recycled ABS, PLA and HIPS thermoplastics for structural applications. PSU Research Review 2018;2(2):115–37.

CHAPTER 7

PVDF-graphene-$BaTiO_3$ composite for 4D applications

Ravinder Sharma[1], Rupinder Singh[2] and Ajay Batish[1]
[1]Department of Mechanical Engineering, Thapar Institute of Engineering and Technology, Patiala, India
[2]Department of Mechanical Engineering, National Institute of Technical Teachers Training and Research, Chandigarh, India

7.1 Introduction

The additive manufacturing (AM)/3D printing techniques has got attention from various disciplines for batch production process [1–3]. The classification of AM process is mainly on the basis of liquid, solid, and powder type raw material [4]. In most commonly used 3D printing processes, semimolten raw material has been deposited in the form of layers over previously deposited material and process remains same until completion of part [5–8]. AM process is completely controlled process, in which initially a program has been saved in the form of G-code that gives command to 3D printer for fabrication of part [9]. Earlier, only thermoplastic based polymers were used in 3D printing process; however, now a days 3D printing of metals, smart materials, and composites of polymers and metal powder has been also processed with AM technologies [10,11]. Fused deposition modeling (FDM) is one of the most widely used 3D printing technique. A thermoplastic polymers or composites based on these polymers have been used in the form of feedstock filament to fabricate a final product [12–14]. In recent years, smart polymer-based composites have been used in 3D printing process [15]. These types of materials transform their shape or other properties in a preprogrammed manner under the effect of external stimulus.

The four-dimensional (4D) printing is one step ahead of 3D printing process in which time is considered as fourth dimension.4D printing incorporates the same 3D printing techniques for fabrication of functional prototypes; however, these 3D printed parts transformed their shapes or other physical properties over time [16,17]. The materials used for 4D printing are also known as smart materials. These materials transforms their shapes in a preprogrammed manner when comes in contact with

4D Printing
DOI: https://doi.org/10.1016/B978-0-12-823725-0.00001-1

heat, humidity, electric field, magnetic field, mechanical forces, etc. [18–22]. The 4D printing is mainly classified into two following types:
- Autonomous 4D printing: In this type of 4D printing, there is no need of any external intervention for transformation in the material. These type of materials react automatically with the change in atmospheric temperature, humidity, etc.
- Nonautonomous 4D printing: In nonautonomous 4D printing, some external trigger is required to charge the material for transformation process. For examples, by applying an electric field, magnetic field, or mechanical force, etc., the materials will alter its properties over time [23–27].

Thus integration of smart polymers at the starting stage of 3D printing for fabrication of structures is also known as 4D printing. Smart materials is one of the most attractive research areas in academic and industrial research [28,29].

Shape memory polymers, electroactive polymers (EAP), and other smart polymers are capable to change/alter their dimensional or physical properties over some period of time when subjected by some external stimuli [30–33]. EAP or piezoelectric polymers produces electricity when comes under mechanical stress, or vice versa of it is known as reverse piezoelectricity [34]. EAPs are generally used in sensor applications and can be easily molded into any shape with integration of AM [35].

For this research work, a literature analysis based on the Web of science database has been done using VOSviewer software package. To obtained the results, three keywords, polyvinylidene fluoride (PVDF) polymer, Piezoelectric property, and AM, have been used. A total of 119 research papers/articles/chapters have been found and downloaded in the ".ris" format, by using "VOSviewer software (version 1.6.16"). Total 33,810 terms have been found for this bibliographic analysis. By selecting minimum occurrences of 6, a total of 88 have met the threshold level. Further from 88 items, best suitable 53 terms have been suggested by VOSviewer software and lastly 40 terms have been selected for mapping and relations among those have been developed. All 40 terms have been divided into four clusters with different colors (red, green, blue, and yellow). Table 7.1 shows the selected term, the number of occurrences within the title + abstract, and relevance score as per the similarity of the entered keywords.

Fig. 7.1 shows the bibliographic relevance analysis of the PVDF-based composites by VOSviewer (based upon Table 7.1). It should be noted that the size and color of node denote the no. of occurrences of term and cluster

Table 7.1 Selected terms for bibliometric analysis with their occurrences and relevance score.

S. No.	Term	Occurrences	Relevance score
1	3D printing	17	0.2324
2	Addition	9	0.1906
3	Additive manufacturing	5	1.083
4	Analysis	8	1.1752
5	Application	23	0.273
6	Barium titanate	6	1.0262
7	Beta phase	6	1.0362
8	Composite	12	1.1571
9	Degrees °C	6	2.4825
10	Deposition modeling	6	0.8938
11	Device	19	0.689
12	Effect	9	1.3338
13	Electric field	10	0.6608
14	Energy harvesting	6	1.4544
15	Fabrication	10	0.2517
16	FDM	6	0.8938
17	Frequency	5	3.8871
18	Low cost	6	1.8805
19	Order	6	0.7963
20	Paper	13	0.7579
21	pC/N	12	1.4557
22	Piezoelectric coefficient	7	1.398
23	Piezoelectric property	13	0.4256
24	Poling	9	0.4439
25	Poly	7	2.2311
26	Polymer	17	0.262
27	Polyvinylidene fluoride	22	0.5118
28	Printing	12	0.5613
29	Process	13	0.4084
30	PVDF	38	0.2411
31	PVDF film	8	0.4282
32	Self	8	0.868
33	Sensing	9	0.7545
34	Sensor	24	0.3718
35	Study	11	1.7614
36	Technique	18	0.381
37	Temperature	13	0.322
38	Time	7	0.7424
39	Vinylidene fluoride	5	3.5528
40	Work	12	0.7238

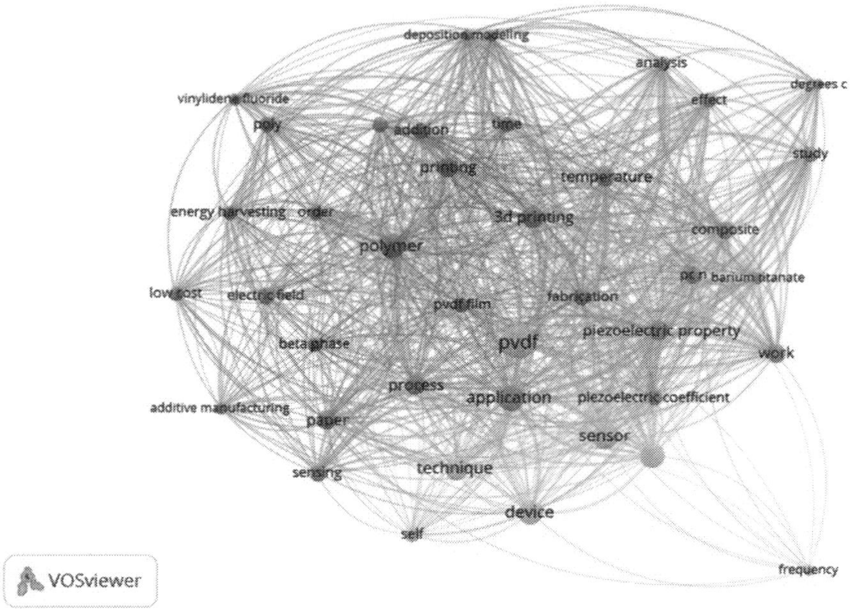

Figure 7.1 Bibliographic map analysis or relationship between all selected 40 terms.

formation of similar studies. Two major clusters have been formed for previously reported studies (represented in red and green color). As observed from Fig. 7.1, number of studies has been reported for the PVDF as a sensor material for use in energy harvesting applications. Researchers have developed the composites based on the PVDF by using different processes. Moreover, 3D printing of polymer matrix for the preparation of the functional/nonfunctional parts in the electric field has also been reported.

Fig. 7.1 shows the bibliographic map analysis or relationship between all selected 40 terms. By clicking upon the node of AM it shows the previously reported work performed in this field. Similar to AM by clicking on the node of PVDF polymer and piezoelectric property the previously reported work conducted over, it is shown in Fig. 7.2A–C.

From Fig. 7.2A–C, it has been ascertained that the studies may be performed for the fabrication of piezoelectric devices by 3D printing of $BaTiO_3$ and Gr as reinforcement in PVDF matrix for 4D applications.

Literature review revealed that AM has been widely used in many fields for rapid production of parts/structures using various types of thermoplastic polymers. But hitherto little has been reported about fabrication of smart polymers-based functional prototypes using FDM. In this

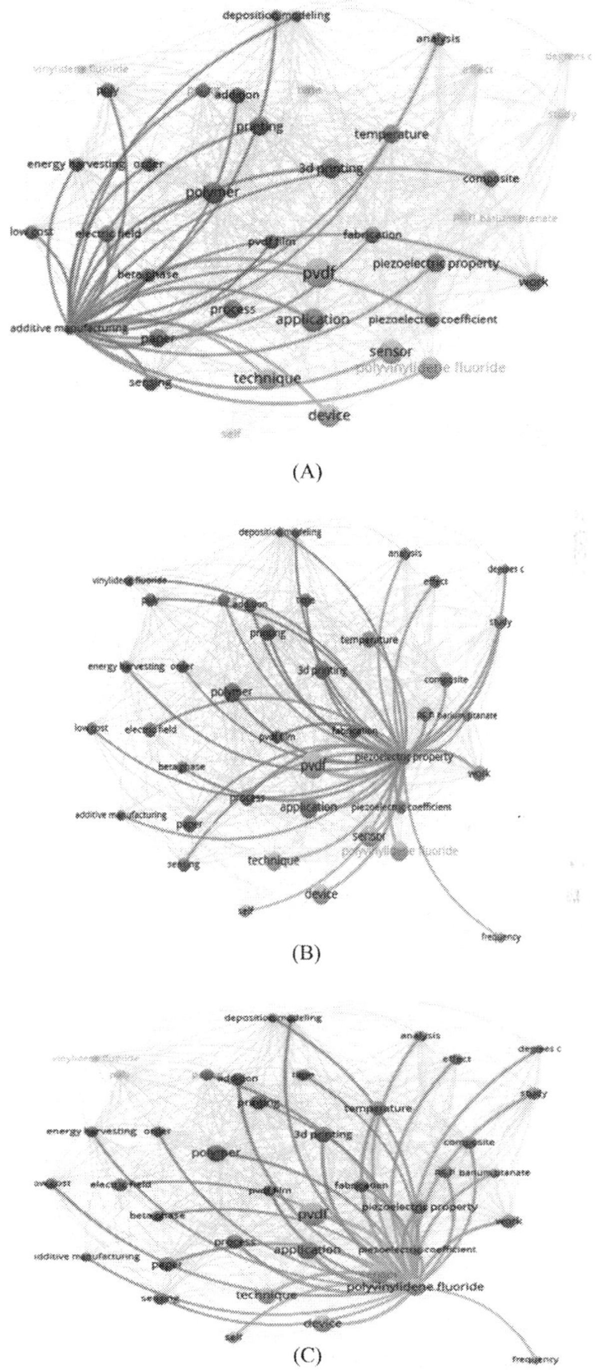

Figure 7.2 Research gaps by selecting node as additive manufacturing (A), piezoelectric property (B), and PVDF (C).

research work, feedstock filament comprising of PVDF, Gr, and piezoceramic $BaTiO_3$ was blended chemically with assistance of mechanical twin screw extruder (TSE). Further, fabricated filament was allowed to run on FDM (an open source 3D printer) setup for fabrication of standard specimens. FDM process parameters have been optimized for surface hardness (SH) of 3D printed parts. The piezoelectric coefficient (D_{33}) has been measured for possible 4D applications.

7.1.1 Experimentation

Methodology adopted for this research work is shown in Fig. 7.3.

PVDF is semicrystalline in nature having polymeric chain of $[-CF_2-CH_2-]_n$. In this research study, solef PVDF was purchased from local vendor (Deval Enterprises, Gujarat, India). The $BaTiO_3$ is white ceramic powder having extraordinary piezoelectric properties. Thus to increase the piezoelectric coefficient of base matrix, $BaTiO_3$ having 100 nm particle size, was reinforced in it. The Gr nano-powder having highly conductive nature was also blended in PVDF matrix. For the preparation of composite, a chemical mixing method was used. At first stage, PVDF was dissolved in dimethyl formamide (DMF) using hot plate magnetic stirrer. The Gr nano-powder and $BaTiO_3$ nano-particles were ultrasonicated in DMF at temperature of 50°C for 30 min. Both the solutions were mixed and further stirred for next 30 min. The slurry developed through this method was poured over the glass substrate and placed inside

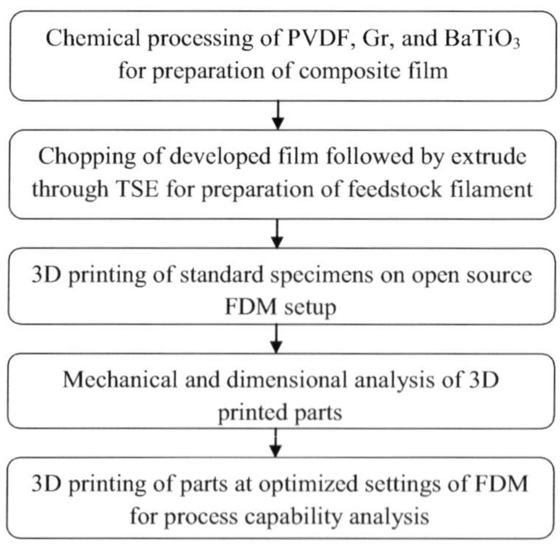

Figure 7.3 Methodology adopted for this research work.

the hot oven at 120°C for 12 hours. The setup used in chemical mixing and developed film of composite is shown in Fig. 7.4.

For preparation of feedstock filament, TSE was used. The thin films prepared from chemical mixing were chopped and fed into the TSE (see Fig. 7.5). The prepared feedstock filament of PVDF/Gr/$BaTiO_3$ was further used to operate on open source FDM without any alteration in software/hardware.

In previously reported studies, the composite of PVDF (83%) + $BaTiO_3$ (15%) + Gr(2%) by weight has shown maximum mechanical properties

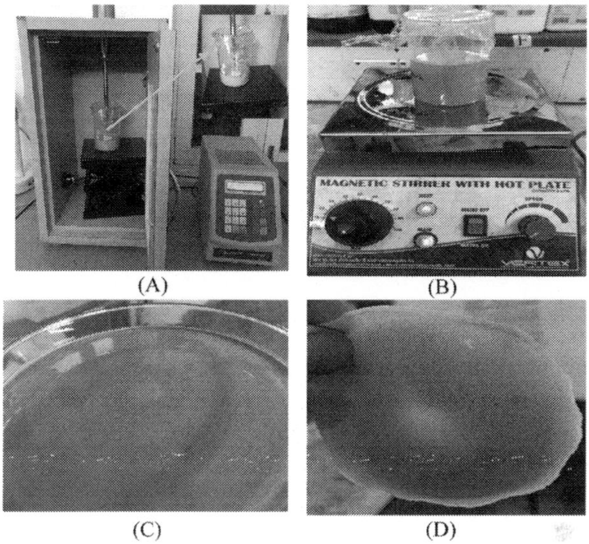

Figure 7.4 Setup used in chemical mixing and developed composite film: (A) ultrasonication process; (B) stirring of solution; (C) slurry poured over glass substrate for heating; (D) thin composite film after evaporation of DMF.

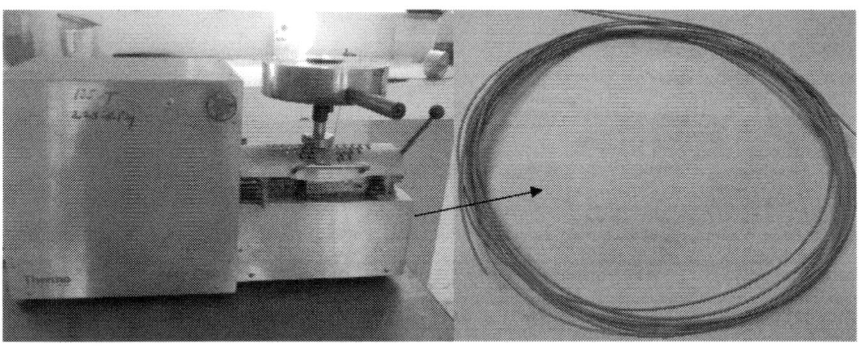

Figure 7.5 TSE used for preparation of feedstock filament and prepared filaments.

when extruded at 180°C screw temperature and 60 rpm of screw speed [32]. Thus similar parameters were used for extrusion of feedstock filament. Further standard tensile specimens were 3D printed using this extruded filament. The selected parameters of FDM for fabrication of standard specimens are shown in Table 7.2.

To minimize the experimental error and enhance the accuracy level, complete design of experimentation was prepared as per Taguchi L9 orthogonal array as shown in Table 7.3.

For 3D printing process an existing FDM setup (Make Divide by zero) was used. Three controllable parameters of FDM were primarily selected such as infill speed (IS), infill angle (IA), and infill density (ID) at three different levels of each (as shown in Table 7.2). Remaining process parameters were kept fixed to evaluate the mechanical properties. SH was measured with shore-D hardness tester. The hardness of fabricated parts was lying in between 65 and 77 shore-D. Table 7.4 shows the outcome values of hardness of fabricated parts.

Table 7.2 Process parameters of FDM (each at three levels).

IS (mm/s)	IA (°)	ID (%)
1. 50	1. 0	1. 60
2. 70	2. 45	2. 80
3. 90	3. 90	3. 100

Table 7.3 Experimental design based on Taguchi L9 OA.

Experiment 1		Experiment 2		Experiment 3	
IS	50	IS	50	IS	50
IA	0	IA	45	IA	90
ID	60	ID	80	ID	100
Experiment 4		Experiment 5		Experiment 6	
IS	70	IS	70	IS	70
IA	0	IA	45	IA	90
ID	80	ID	100	ID	60
Experiment 7		Experiment 8		Experiment 9	
IS	90	IS	90	IS	90
IA	0	IA	45	IA	90
ID	100	ID	60	ID	80

7.2 Results and discussion

SH of the 3D printed specimens were measured and found that experiment performed with minimum level of IS 50 mm/s and IA 90 degrees with 100% ID has shown best SH (Table 7.4). The analysis of variance (ANOVA) table for SH is shown in Table 7.5. As observed from Table 7.5, only one parameter was found significant having *P* value less than .05, whereas other two parameters IA and ID were not significant at 95% confidence level, because *F* value is not more than 20. The residual error is found less than 4% represents that model was significant.

Based upon Table 7.5, Table 7.6 shows the rank of input variable at the condition of higher the better for observed SN (signal-to-noise) values of SH. It has been observed from the rank table that IS played most prominent role in shore-D values of fabricated parts, whereas ID and IA stood second and third, respectively. The IA has not shown much effect on SH.

Table 7.4 Surface hardness (shore-D) values of fabricated parts.

Experiment No.	IS (mm/s)	IA (°)	ID (%)	Shore-D hardness
1	50	0	60	73.25
2	50	45	80	74.48
3	50	90	100	76.89
4	70	0	60	71.57
5	70	45	80	72.57
6	70	90	100	69.84
7	90	0	60	72.04
8	90	45	80	65.07
9	90	90	100	66.34

Table 7.5 ANOVA table for surface hardness.

Factor	DoF	Seq. SS	Adj SS	Adj MS	*F*	*P*	% Age contribution
IS	2	1.127	1.127	0.563	22.45	.043	65.67
IA	2	0.073	0.073	0.036	1.44	.409	4.2
ID	2	0.466	0.466	0.233	9.29	.097	27.15
Residual error	2	0.052	0.052	0.025			3.03
Total	8	1.716					

Adj SS, Adjusted sum of squares; *DoF*, degree of freedom; *F*, Fisher's value; *P*, probability; *Seq SS*, sum of squares.

Table 7.6 Rank table of variable of FDM.

Levels	IS	IA	ID
1	37.48	37.18	36.82
2	37.06	36.97	36.99
2	36.62	37.01	37.36
Delta	0.87	0.21	0.56
Rank	1	3	2

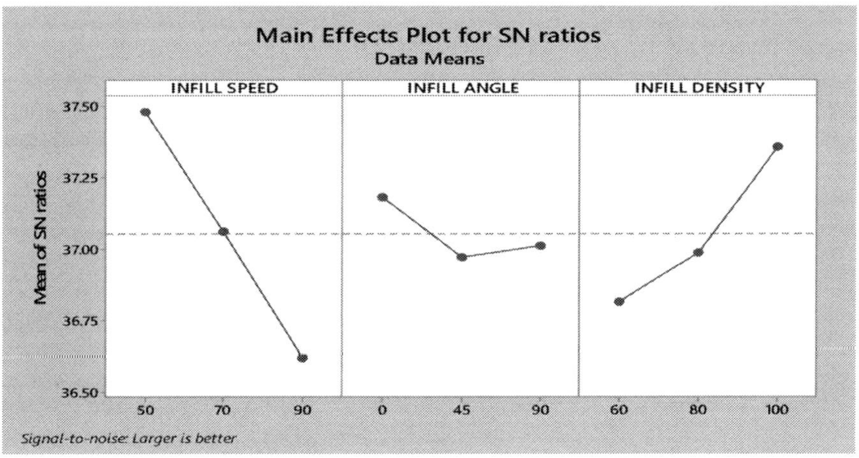

Figure 7.6 Graphical representation of SN ratios.

Fig. 7.6 shows the plot of S/N values for SH. The ID at level first (50 mm/s), first level of IA (0 degree), and third level of ID (100%) have been found as best process variables settings of FDM for manufacturing of PVDF + Gr + $BaTiO_3$ composite.

Further, the SN values observed have been used to calculate the optimum values of shore-D hardness at higher the better status by using below mentioned equations.

$$\eta_{\text{opt}} = G + (G_A - G) + (G_B - G) + (G_C - G) \tag{7.1}$$

$$Z_{\text{opt}}{}^2 = \left(1/10\right)^{\eta_{\text{opt}}/10} \tag{7.2}$$

$$y_{\text{opt}}{}^2 = (10)^{\eta_{\text{opt}}/10} \tag{7.3}$$

Eqs. (7.2) and (7.3) are mainly used for lesser the better and higher the better, respectively. So, to calculate the optimized value of SH, following equation has been used:

$\eta_{(opt)} = G + (G_A - G) + (G_B - G) + (G_C - G)$

${Y_{opt}}^2 = (10)^{\eta_{opt}/10}$ for properties where, largest is better

G = Mean of SN for SH = 37.05

G_A = Max IS from rank table = 37.48

G_B = Max IA from rank table = 37.18

G_C = Max ID from rank table = 37.36

$$\eta_{(opt)} = 37.05 + (37.48 - 37.05) + (37.18 - 37.05) + (37.36 - 37.05)$$

$$\eta_{(opt)} = 37.92$$

$$Y_{opt}^2 = (10)^{37.92/10}$$

$$Y_{opt} = 78.70 \ \text{Shore-}D$$

The optimized value of SH is 78.70 shore-D. Further three confirmatory experiments were conducted at the optimum conditions (i.e., IS 50 mm/s, IA 0 degree, and ID 100%). The experimental observed value of SH is found very close to calculated value, that is, 78.50 shore-D.

7.2.1 Dimensional analysis

After the SH testing, manufactured specimens were subjected to dimensional analysis. A digital Vernier calliper having accuracy up to two decimal places was used to measure the thickness of specimens. To minimize the measurement error, each sample was measured three times. The average thickness of measured dimensions of the 3D printed specimens was compared with the standard dimensions of tensile specimens (as shown in Table 7.7). It has been found that the parts manufactured/fabricated at IS of 70 mm/s, and depositing material at an IA 90 degrees, with 100% ID shown least deviation. Whereas, 3D printing performed as per experiment setting having IS 90 mm/s with IA 90 degrees and 100% ID has shown maximum deviation.

Finally, 10 standard tensile specimens were 3D printed at optimized settings of FDM (IS 50 mm/s, IA 0 degree, with 100% ID) to report the process capability analysis of 3D printing process parameters. The 3D printed functional prototypes were subjected to shore-D hardness testing and dimensional analysis. The output results of mechanical testing of 3D printed specimens in the form of shore-D values and dimensional deviation were noted to perform process capability analysis (Table 7.8).

Based upon Table 7.8, C_p and C_{pk} values were calculated (Table 7.9). The outcome graphs of process capability analysis in the form of histogram and normal probability plot for shore-D hardness and dimensional deviation of 3D printed specimen are shown in Fig. 7.7. The histogram for the observed values of shore-D hardness shown that values were inside the upper statistical limits (USL) and lower statistical limits (LSL) and the bell-shaped curve outlined that the observed result followed the normal distribution. Further, in normal probability plot, all the observed values of peak strength were lies near the normal line, which outlined that process is in control.

Table 7.7 Dimensional deviation observed from dimensional analysis.

Exp. No.	IS (mm/s)	IA (°)	ID (%)	T1	T2	T3	T avg	T (required)	ΔT
1	50	0	60	2.91	2.99	3.04	2.98	3.2	0.22
2	50	45	80	2.98	3.05	3.12	3.05	3.2	0.15
3	50	90	100	3.10	3.12	3.20	3.14	3.2	0.06
4	70	0	60	3.10	3.15	3.19	3.14	3.2	0.05
5	70	45	80	3.05	3.01	2.97	3.01	3.2	0.19
6	70	90	100	3.19	3.0	3.26	3.15	3.2	0.05
7	90	0	60	3.21	3.05	3.04	3.10	3.2	0.10
8	90	45	80	2.99	2.85	3.10	2.98	3.2	0.22
9	90	90	100	2.80	3.10	2.83	2.91	3.2	0.29

Table 7.8 Shore-D hardness testing and dimensional analysis of 3D printed specimens.

S. No.	Shore-D hardness	Dimensional deviation
1	78.50	0.10
2	78.20	0.05
3	78.78	0.08
4	78.69	0.10
5	78.85	0.04
6	78.48	0.08
7	78.34	0.09
8	78.74	0.05
9	78.90	0.06
10	78.28	0.04

Table 7.9 Results of process capability analysis for 3D printed specimens.

Std. deviation	Shore-D hardness 0.22742	Dimensional deviation 0.019493
C_p	1.17	1.28
C_{pk}	1.06	1.18

(1) C_p and C_{pk} measures consistency with average performance. The "k" stands for "centralizing factor." The index takes into consideration the fact that data is maybe not centered. (2) For 3D printed specimen: (a) For shore-D hardness, the USL and LSL were 79.3 and 77.7, respectively; (b) For dimensional deviation, the USL and LSL were 0.15 and 0.00, respectively.

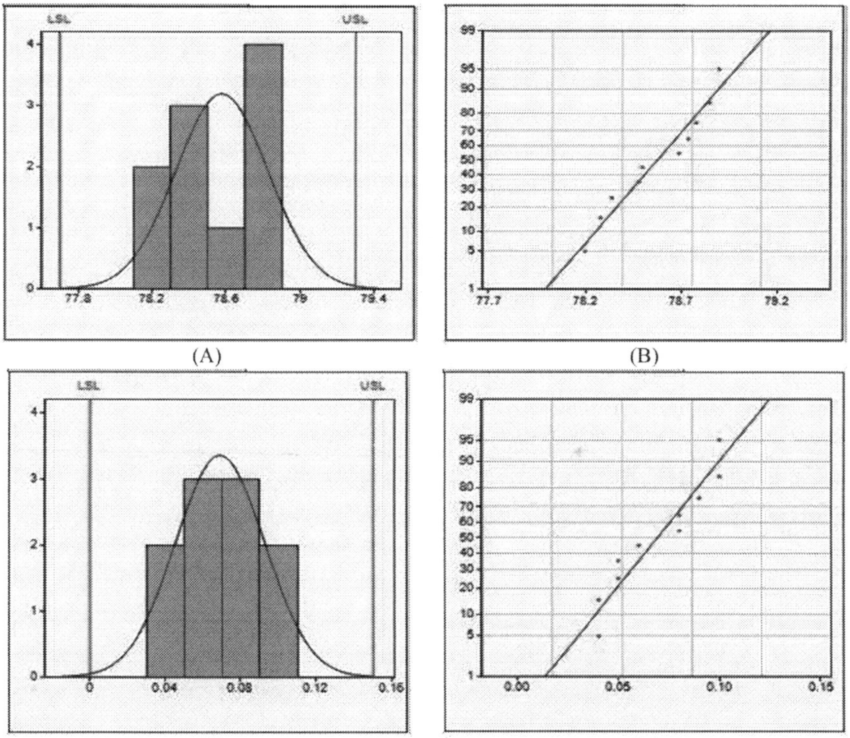

Figure 7.7 Histogram and normal distribution plot obtained from process capability analysis. (A) and (C) Hisogram for Shore-D hardness and dimensionall deviation, (B) and (D) Normal probability plot for Shore-D and dimensional deviation obtained from process capability analysis.

7.2.2 3D printing of piezoelectric sensor

The optimized settings of FDM were used to 3D printing of thin cylinder disk for evaluation of 4D properties of developed composites. The diameter of disk was 10 mm with 0.4 mm thickness. The 3D printing was performed by keeping the nozzle temperature 260°C and bed temperature was kept 80°C. Material deposition rate was kept high to avoid the burn and clog of composite material inside the nozzle opening. After successfully preparation of thin cylindrical disks, silver paint was coated on both surfaces of the disk to make it conductive. As the electric poling was performed at high temperature, coating was done with temperature resistive silver paint. Thus for conversion of PVDF β-phase, high voltage electric field was applied through the disk. To avoid any type of breakdown at high voltage, sample was dipped in silicon oil during the poling process. The temperature was kept at 120°C during the electrical poling process. The complete setup of DC poling unit is shown in Fig. 7.8.

To measure the piezoelectric coefficient of the PVDF/Gr/$BaTiO_3$-based 3D printed film a D_{33} meter (Model YE2730A D33, purchased from Marine India a local manufacturer of Delhi, India) having resolution of 0.1 pC/N was used in this research work The piezoelectric coefficient of the piezo ceramics, polymers, and piezoelectric crystals can be directly measure by this meter. After electrical poling, the disk was placed over the lower electrode or in between the two probes of D_{33} meter. Fig. 7.9 shows the D_{33} measuring meter with specimen to measure the piezoelectric coefficient, and probes holding the silver painted fabricated disk. The meter is then switched on, that displays the digital value of D_{33} in pC/N. The piezoelectric constant (D_{33}) value of 3D printed disk is 30.2 pC/N.

Figure 7.8 Setup used for electrical poling of specimen.

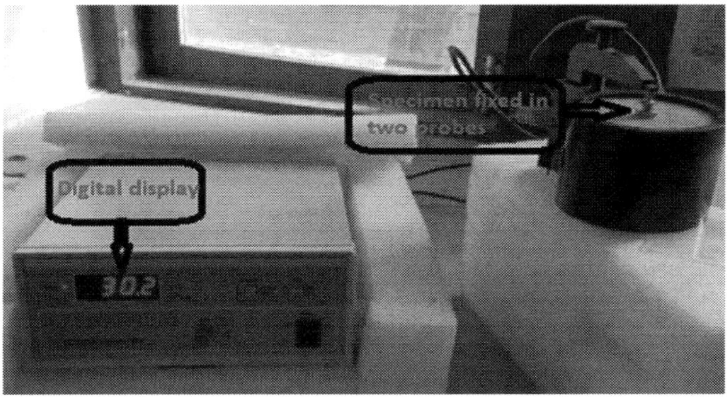

Figure 7.9 Setup used to measure the piezoelectric coefficient (D_{33}).

7.3 Conclusion

This study reports fabrication of smart composite-based feedstock filament of PVDF-based polymer doped with electrical conductive Gr powder and piezo-ceramic $BaTiO_3$. Further, the in-house developed feedstock filament was used on FDM machine for manufacturing of standard specimens. Following are the conclusions from this study:

- The results of ANOVA highlighted that IS played a dominated role toward the hardness of fabricated parts surfaces (i.e., 65.67%) followed by ID that affected only 27%. However, IA was found not significant with P value more than .05. The results of dimensional analysis suggested that IS 70 mm/s, IA 90 degree, and ID 100% provide minimum dimensional deviation.
- The values of C_p and C_{pk} were found more than 1 from process capability analysis represents that 3D printing of specimens at FDM is statistically under control process for batch production applications.
- The piezoelectric coefficient (D_{33}) of 3D printed parts was found 30.2 pC/N. This may be used for 4D printing applications (such as self-assembly, controlled volumetric change, etc.).

References

[1] Bundhoo ZM. Renewable energy exploitation in the small island developing state of Mauritius: current practice and future potential. Renewable and Sustainable Energy Reviews 2018;82:2029−38.

[2] Azarpour A, Suhaimi S, Zahedi G, Bahadori A. A review on the drawbacks of renewable energy as a promising energy source of the future. Arabian Journal for Science and Engineering 2013;38(2):317−28.

[3] Anton SR, Sodano HA. A review of power harvesting using piezoelectric materials (2003–2006). Smart materials and Structures 2007;16(3):R1.
[4] Tarelho JP, dos Santos MP, Ferreira JA, Ramos A, Kopyl S, Kim SO, et al. Graphene-based materials and structures for energy harvesting with fluids—A review. Materials Today 2018;21(10):1019–41.
[5] Shepelin NA, Glushenkov AM, Lussini VC, Fox PJ, Dicinoski GW, Shapter JG, et al. New developments in composites, copolymer technologies and processing techniques for flexible fluoropolymer piezoelectric generators for efficient energy harvesting. Energy & Environmental Science 2019;12(4):1143–76.
[6] Bowen CR, Kim HA, Weaver PM, Dunn S. Piezoelectric and ferroelectric materials and structures for energy harvesting applications. Energy & Environmental Science 2014;7(1):25–44.
[7] Sharma R, Singh R, Batish A. Investigations for barium titanate and graphene reinforced PVDF matrix for 4D applications. Encyclopedia of Renewable and Sustainable Materials. ELSEVIER 2020;366–375. Available from: https://doi.org/10.1016/B978-0-12-803581-8.11306-2.
[8] Damjanovic D, Newnham RE. Electrostrictive and piezoelectric materials for actuator applications. Journal of Intelligent Material Systems and Structures 1992;3(2):190–208.
[9] Sabry RS, Hussein AD. PVDF: ZnO/$BaTiO_3$ as high out-put piezoelectric nanogenerator. Polymer Testing 2019;79:106001.
[10] Liu Z, Xing Z, Wang H, Xue Z, Chen S, Jin G, et al. Effect of ZnO on the microstructure and dielectric properties of BaTiO3 ceramic coatings prepared by plasma spraying. Journal of Alloys and Compounds 2017;727:696–705.
[11] He S, Dong W, Guo Y, Guan L, Xiao H, Liu H. Piezoelectric thin film on glass fiber fabric with structural hierarchy: an approach to high-performance, superflexible, cost-effective, and large-scale nanogenerators. Nano Energy 2019;59:745–53.
[12] Kuder IK, Arrieta AF, Raither WE, Ermanni P. Variable stiffness material and structural concepts for morphing applications. Progress in Aerospace Sciences 2013;63:33–55.
[13] Ghosh SK, Adhikary P, Jana S, Biswas A, Sencadas V, Gupta SD, et al. Electrospungelatin nanofiber based self-powered bio-e-skin for health care monitoring. Nano Energy 2017;36:166–75.
[14] Mehrban N, Teoh GZ, Birchall MA. 3D bioprinting for tissue engineering: stem cells in hydrogels. International Journal of Bioprinting 2016;2(1).
[15] Bera B, Sarkar MD. Piezoelectricity in PVDF and PVDF based piezoelectric nanogenerator: a concept. IOSR Journal of Applied Physics 2017;9(3):95–9.
[16] Nalwa HS. Recent developments in ferroelectric polymers. Journal of Macromolecular Science, Part C: Polymer Reviews 1991;31(4):341–432.
[17] Tamang A, Ghosh SK, Garain S, Alam MM, Haeberle J, Henkel K, et al. DNA-assisted β-phase nucleation and alignment of molecular dipoles in pvdf film: a realization of self-poled bioinspired flexible polymer nanogenerator for portable electronic devices. ACS Applied Materials & Interfaces 2015;7(30):16143–7.
[18] Chen X, Han X, Shen QD. PVDF-based ferroelectric polymers in modern flexible electronics. Advanced Electronic Materials 2017;3(5):1600460.
[19] Sharma R, Singh R, Batish A. On mechanical and surface properties of electroactive polymer matrix-based 3D printed functionally graded prototypes. Journal of Thermoplastic Composite Materials 2020. Available from: https://doi.org/10.1177/0892705720907677.
[20] Sharma R, Singh R, Batish A. Study on barium titanate and graphene reinforced PVDF matrix for 4D applications. Journal of Thermoplastic Composite Materials 2019. Available from: https://doi.org/10.1177/0892705719865004.

[21] Pal A, Sasmal A, Manoj B, Rao DP, Haldar AK, Sen S. Enhancement in energy storage and piezoelectric performance of three phase (PZT/MWCNT/PVDF) composite. Materials Chemistry and Physics 2020;244:122639.
[22] Mangat AS, Singh S, Gupta M, Sharma R. Experimental investigations on natural fiber embedded additive manufacturing-based biodegradable structures for biomedical applications. Rapid Prototyping Journal 2018.
[23] Ranjan N, Singh R, Ahuja IP, Singh J. Fabrication of PLA-HAp-CS based biocompatible and biodegradable feedstock filament using twin screw. Additive Manufacturing of Emerging Materials 2018;325.
[24] Singh S, Ramakrishna S, Singh R. Material issues in additive manufacturing: a review. Journal of Manufacturing Processes 2017;25:185–200.
[25] Bourell DL, Rosen DW, Leu MC. The roadmap for additive manufacturing and its impact. 3D Printing and Additive Manufacturing 2014;1(1):6–9.
[26] Sharma R, Singh R, Penna R, Fraternali F. Investigations for mechanical properties of HAp, PVC and PP based 3D porous structures obtained through biocompatible FDM filaments. Composites Part B: Engineering 2018;132:237–43.
[27] Singh R, Kumar R, Ranjan N, Penna R, Fraternali F. On the recyclability of polyamide for sustainable composite structures in civil engineering. Composite Structures 2018;184:704–13.
[28] Ranjan N, Singh R, Ahuja IP. Development of PLA-HAp-CS-based biocompatible functional prototype: a case study. Journal of Thermoplastic Composite Materials 2020;33(3):305–23.
[29] Choi J, Kwon OC, Jo W, Lee HJ, Moon MW. 4D printing technology: a review. 3D Printing and Additive Manufacturing 2015;2(4):159–67.
[30] Singh R, Sharma R, Davim JP. Mechanical properties of bio compatible functional prototypes for joining applications in clinical dentistry. International Journal of Production Research 2018;56(24):7330–40.
[31] Mohan N, Senthil P, Vinodh S, Jayanth N. A review on composite materials and process parameters optimisation for the fused deposition modelling process. Virtual and Physical Prototyping 2017;12(1):47–59.
[32] Sharma R, Singh R, Batish A. On effect of chemical-assisted mechanical blending of barium titanate and graphene in PVDF for 3D printing applications. Journal of Thermoplastic Composite Materials 2020. Available from: https://doi.org/10.1177/0892705720945377.
[33] Spaggiari A, Castagnetti D, Golinelli N, Dragoni E, ScirèMammano G. Smart materials: properties, design and mechatronic applications. Proceedings of the Institution of Mechanical Engineers, Part L: Journal of Materials: Design and Applications 2019;233(4):734–62.
[34] Yu X, Cheng H, Zhang M, Zhao Y, Qu L, Shi G. Graphene-based smart materials. Nature Reviews Materials 2017;2(9):1–3.
[35] Farina I, Sharma R, Singh R, Batish A, Singh N, Fraternali F, et al. Mechanical characterization of FDM filaments with PVDF matrix reinforced with graphene and barium titanate. IOP Conference Series: Materials Science and Engineering 2020;999 (1):012010.

CHAPTER 8

Hydrothermal stimulus for 4D capabilities of PA6-Al-Al_2O_3 composite

Kamaljit Singh Boparai[1] and Rupinder Singh[2]
[1]Department of Mechanical Engineering, GZSCCET, MRS Punjab Technical University, Bathinda, India
[2]Department of Mechanical Engineering, National Institute of Technical Teachers Training and Research, Chandigarh, India

8.1 Introduction

The term 4D printing was first introduced by Tibbits in 2013 [1]. Since then, 4D printing fetch the attention of researchers toward smart materials. Initially, 4D printing is defined as the combination of 3D printing and time, that is, the shape, property, or functionality of a 3D printed structure can change with respect to time [2]. Momeni et al. briefly defined it as the targeted evolution of the 3D printed structure, in terms of shape, property, and functionality [3]. It has a capability of self-assembly, multifunctionality, and self-repairs. More particularly, it is a time-dependent, printer-independent, and has predictable response [4]. In other worlds, the 3D printed structure when exposed to external stimuli (such as water, heat, light, and their combinations such as water heat, heat light, water light, magnetic field, pH, etc.) results in dynamic structure. Fig. 8.1 shows the concept of 4D printing.

The main pillars of 4D printing are 3D printer, external stimulus, stimulus responsive materials, interaction mechanism, and mathematical modeling [3]. The 3D printer facilitates the printing of 3D structures with a provision of single material or multimaterial printing. The external stimulus causes the change in shaper/property/functionality of 3D printed part. The selection of stimulus depends upon the desired output and response of the feedstock filament material employed in 3D printer. Presently, smart or stimulus responsive materials have been explored by researchers and categorized based upon their distinguish features such as self-assembly, self-repair, shape memory, etc. The studies also reported the use of some interaction mechanism for stimulus and materials to formulate 4D printing capabilities which otherwise

4D Printing
DOI: https://doi.org/10.1016/B978-0-12-823725-0.00005-9

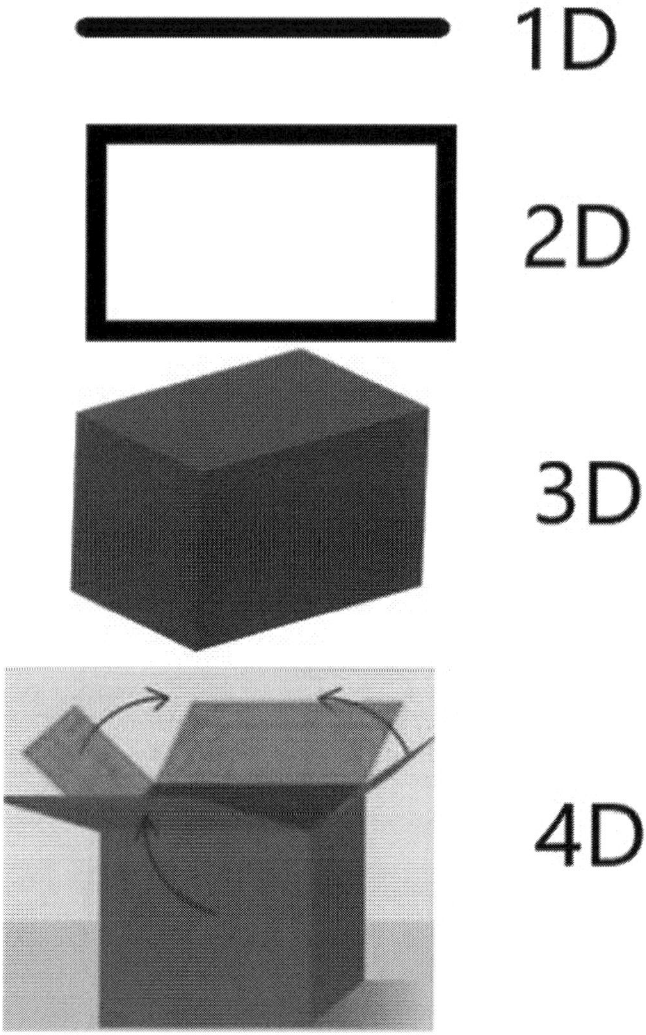

Figure 8.1 4D printing concept.

impossible to achieve. Further, mathematical modeling also assists the design process for material distribution and structure, to get the desired shape, property, and functionality. The 4D printing may be one-way programmed or two-way program (Fig. 8.2).

The most commonly used 3D printing technologies associated with 4D printing are free form fabrication, digital ink writing, inkjet printing, selective laser sintering, digital light processing, and stereo lithography. The literature

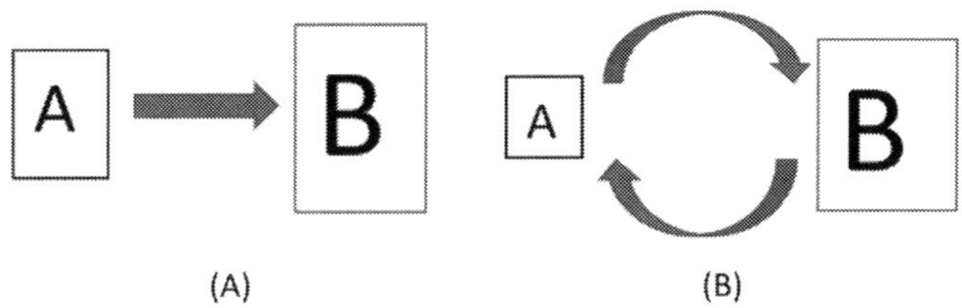

Figure 8.2 4D printing; (A) one way programmed; (B) two way programmed.

also highlighted the use of responsive materials such as shape change materials (SCM) [5–7]. The SCM are further classified as shape memory polymers [8,9], shape memory alloys [10–12], shape memory ceramics [13–15], shape memory gels [16–18], and shape memory hybrid [19–21].

In order to ascertain the research gap, web of science data base of past 20 years was explored (2001–20). Initially the search was performed with keyword "4D printing" and 756 results were obtained from core collection of web of science database. Out of these results, recent 500 articles were selected for further analysis. The VOS viewer (open-source software) was used for bibliographic analysis and preparation of networking diagrams. For the selected keyword by fixing minimum number of occurrences as "5," out of 11941 terms, 588 meet the threshold. For each of 588 terms, relevance score was calculated and further 60% most relevant terms, that is, 353 were used for analysis. Fig. 8.3 shows networking diagram for keyword 4D printing. As observed from Fig. 8.3, 04 clusters were noticed which represents the major work reported in particular application domain.

Further to ascertain the research gap, Fig. 8.4A–C shows research gap analysis by selecting nodal points as review, fused deposition modeling (FDM), and experiment.

After this, the web of science data base was explored for keyword "4D capabilities" and 689 results were obtained. For the selected keyword by fixing minimum number of occurrences as "5," out of 16335 terms, 515 meet the threshold. For each of 515 terms, relevance score was calculated and further 60% most relevant terms (309) were used for analysis. Fig. 8.5 shows networking diagram for keyword "4D capabilities." As observed from Fig. 8.5, 04 clusters were noticed which represents the major work reported in particular application domain.

Further to ascertain the research gap, Fig. 8.6A–C shows research gap analysis by selecting nodal points as 4D printing, structures, and mechanism.

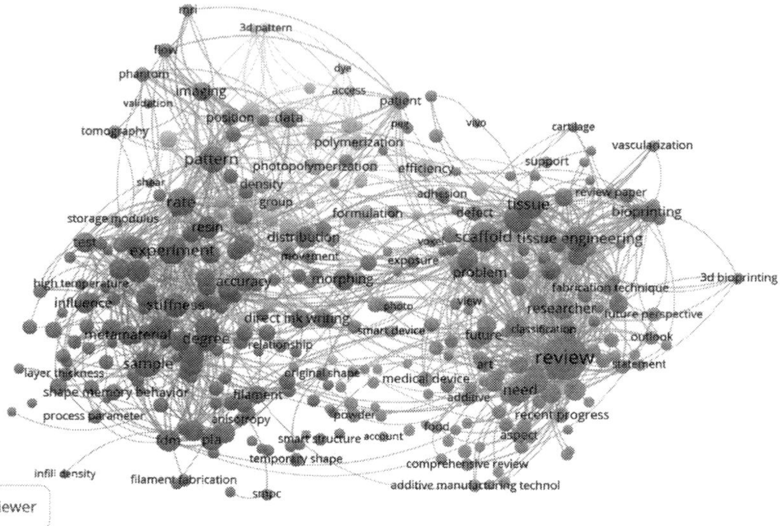

Figure 8.3 Networking diagram for keyword 4D printing.

Finally, the keyword "stimulus in 4D printing" was used and total 208 results were obtained in core selection of web of science data. For the selected keyword by fixing minimum number of occurrences as "5," out of 5089 terms, 259 meet the threshold. For each of 259 terms, relevance score was calculated and further 60% most relevant terms (155) were used for analysis (Table 8.1). Fig. 8.7 shows networking diagram for keyword "stimulus in 4D printing." As observed from Fig. 8.7, 03 clusters were noticed which represents the major work reported in particular application domain.

Further to ascertain the research gap, Fig. 8.8A–C shows research gap analysis by selecting nodal points as manufacturing, fourth dimension, and shape memory.

The web of science-based research gap analysis highlighted that the limited work has been done on the investigations of 4D capabilities of composite materials with hydrothermal stimulus. The purpose of this research work is to understand the 4D capabilities of polyamide (PA)6-Al-Al_2O_3-based composite feedstock filaments for FDM process. The specimens have been exposed to different hydrothermal stimuli conditions and the change in dimensions was observed before and after stretching under controlled conditions (30 mm/minute for 10 seconds) on tensile testing machine. The results are also compared with the acrylonitrile butadiene styrene (ABS), standard feedstock filament. The following methodology has been used for this research work. Fig. 8.9 shows the process flow diagram.

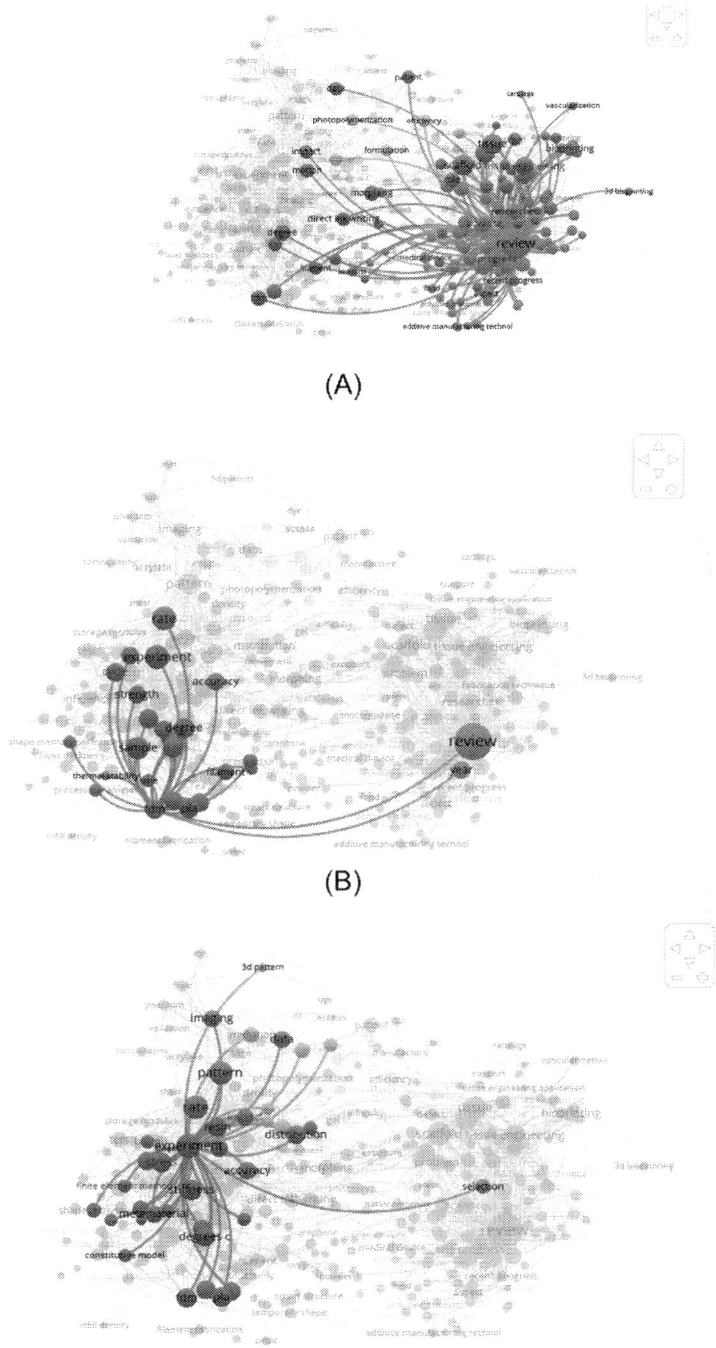

(C)

Figure 8.4 Gap analysis by selecting nodal points as (A) review, (B) fused deposition modeling, and (C) experiment.

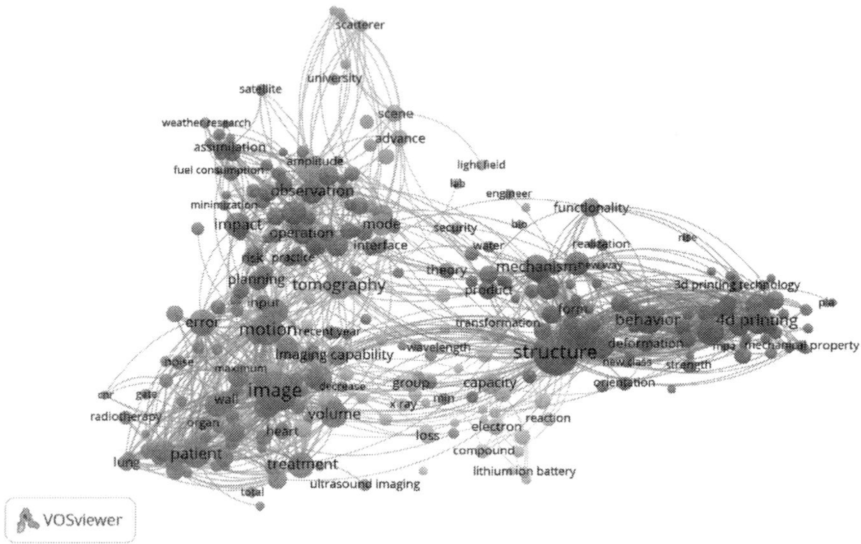

Figure 8.5 Networking diagram for "4D capabilities."

8.2 Experimentation

8.2.1 Materials

As already mentioned, PA 6-based composite feedstock filaments, reinforced with Al and Al_2O_3, for FDM process, have been prepared on single screw extruder, by operating at optimum combination of critical process parameters [22,23]. The weight proportions of the compositions considered for experimentations are reported in Table 8.2.

8.2.2 Sample preparation

The feedstock filaments (with different composition/ proportion) prepared on single screw extruder has Ø1.75 ± 0.05 mm. For hydrothermal analysis, the length of the feedstock filaments has been taken as 200 mm. Fig. 8.10 shows the specimens for hydrothermal testing.

8.2.3 Sample processing

As already mentioned above that the hydrothermal behavior of PA6-Al-Al_2O_3-based composite feedstock filaments have been analyzed by stretching the specimens (having diameter 1.75 mm and gage length 100 mm) with a controlled pulling speed of 30 mm/minute for 10 seconds. Four different conditions have been created;

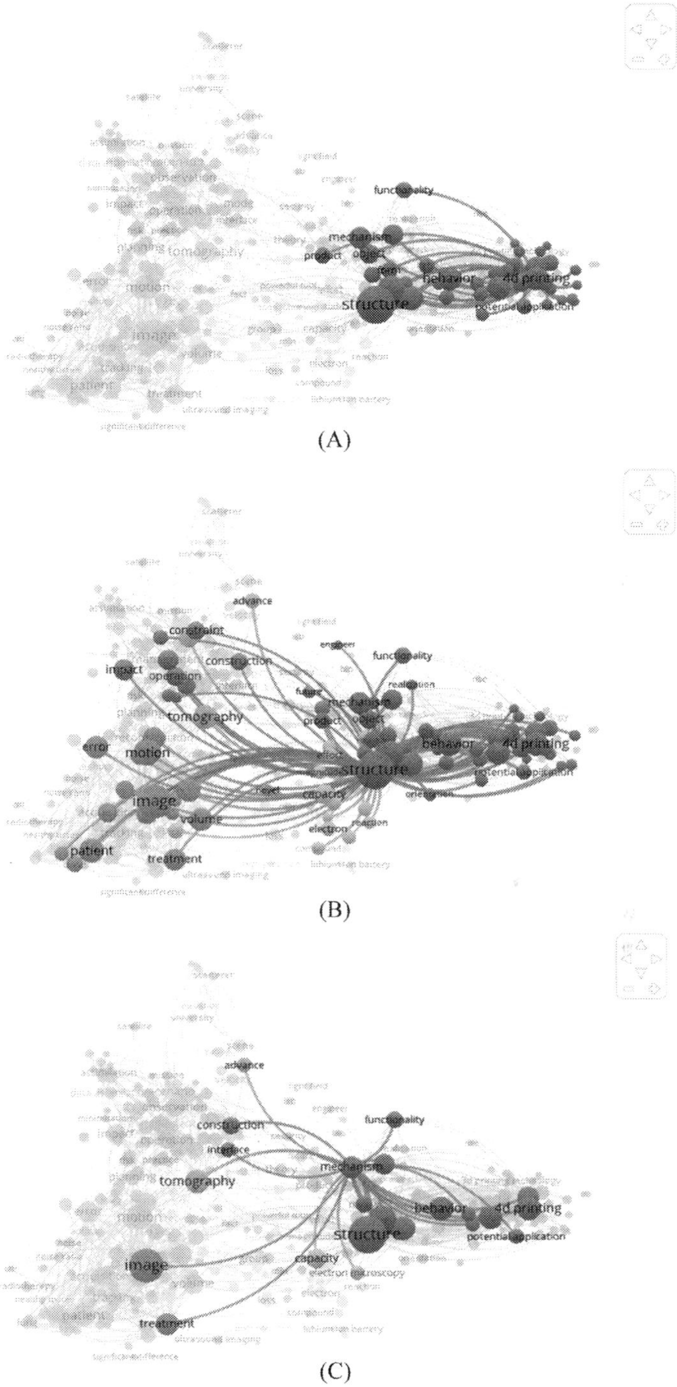

Figure 8.6 Gap analysis by selecting nodal points as (A) 4D printing, (B) structures, and (C) mechanism.

Table 8.1 Relevance score calculation for keyword "stimulus in 4D printing."

Id	Term	Occurrences	Relevance score
1	3D bioprinting	6	4.7434
2	3D printer	8	0.8839
3	4D bioprinting	8	3.9173
4	4d printed object	5	0.8629
5	4D printed structure	7	0.607
6	4D printing process	7	0.7116
7	4D printing technology	15	0.6198
8	Fourth dimension	6	1.0132
9	Active material	11	0.4756
10	Additive manufacturing technique	6	1.1232
11	Additive manufacturing technology	5	1.053
12	Advance	14	0.5456
13	Aerospace	7	0.709
14	Architecture	12	0.2951
15	Art	8	1.1185
16	Aspect	13	0.5613
17	Attention	10	0.6267
18	Benefit	7	1.9678
19	Bio	5	1.1117
20	Biocompatibility	7	1.4037
21	Biomedical application	9	0.7981
22	Biomedical device	12	0.4635
23	Biomedical engineering	5	0.9165
24	Biomedical field	8	1.3889
25	Biomedicine	5	0.775
26	Bioprinting	9	4.5751
27	Cell	13	1.7501
28	Chemistry	8	0.9205
29	Class	15	0.6038
30	Color	9	0.8206
31	Composition	8	0.747
32	Comprehensive review	5	1.3563
33	Configuration	14	0.4337
34	Construct	11	3.0006
35	Creation	7	0.7037
36	Current challenge	6	1.258
37	Customization	7	0.7452
38	Deformation	32	0.4947
39	Degree	15	0.4655
40	Degrees c	11	0.758
41	Direct ink writing	9	0.8691

(Continued)

Table 8.1 (Continued)

Id	Term	Occurrences	Relevance score
42	Distribution	14	0.4381
43	Digital ink writing	5	1.3773
44	Drug delivery	11	0.6935
45	Engineering	11	0.4964
46	Environment	14	0.8241
47	Evolution	11	0.8237
48	Example	13	0.453
49	Experiment	12	0.5272
50	Extrusion	10	0.9679
51	Fabrication method	6	1.4588
52	Fused deposition modeling	5	1.5855
53	Feasibility	8	0.8425
54	Flexibility	7	0.7166
55	Flower	8	0.4753
56	Focus	10	0.738
57	Force	15	0.4233
58	Formation	7	1.5583
59	Formulation	8	1.1895
60	Fourth dimension	16	0.6431
61	Framework	10	0.6252
62	Functional material	8	0.9963
63	Future	7	0.7628
64	Future application	7	1.5679
65	Future perspective	9	3.7074
66	Gel	9	1.1471
67	Heating	16	0.7059
68	Humidity	12	0.4297
69	Improvement	7	0.7365
70	Influence	8	0.5824
71	Ink	18	0.7166
72	Intelligent material	7	0.6705
73	Investigation	8	0.8394
74	Lack	6	0.7645
75	Large deformation	5	1.35
76	Lce	6	1.7755
77	Limitation	12	0.5163
78	Liquid crystal elastomer	7	1.405
79	Low cost	6	1.6265
80	Manufacturing	55	0.2468
81	Medicine	7	1.0382
82	Metal	6	1.0912

(Continued)

Table 8.1 (Continued)

Id	Term	Occurrences	Relevance score
83	Modulus	10	1.1886
84	Morphing	7	0.9117
85	Motion	12	0.4574
86	Multimaterial	6	1.8107
87	Multiple material	5	0.9717
88	N isopropylacrylamide	9	0.9808
89	New technology	5	1.8918
90	Number	8	0.8558
91	Organ	7	2.1691
92	Original shape	7	0.8117
93	Outlook	8	0.7149
94	Overview	10	0.7049
95	Parameter	20	0.6935
96	Past decade	5	1.1634
97	Pattern	22	0.3427
98	Permanent shape	6	1.6796
99	Pla	9	1.9513
100	Poly	13	0.7052
101	Polylactic acid	10	1.8309
102	Practical application	9	0.6564
103	Pre	6	0.5195
104	Presence	6	1.0192
105	Print	8	0.8712
106	Printed object	8	0.58
107	Printer	11	0.6086
108	Printing method	10	0.3315
109	Printing process	11	0.6668
110	Problem	11	0.4916
111	Range	8	1.1735
112	Rapid prototyping	7	1.3564
113	Ratio	15	0.7542
114	Recent advance	9	0.8349
115	Recent year	8	0.8847
116	Researcher	8	0.9825
117	Responsive hydrogel	7	0.849
118	Rheological property	5	1.0738
119	Robotic	33	0.242
120	Sample	9	1.2294
121	Scaffold	13	0.8937
122	Self-assembly	5	2.5284
123	Self-healing	7	0.8767

(Continued)

Table 8.1 (Continued)

Id	Term	Occurrences	Relevance score
124	Shape change	24	0.3876
125	Shape memory	6	1.216
126	Shape memory effect	13	0.6387
127	Shape memory material	8	0.8187
128	Shape morphing	8	0.7521
129	Shape recovery	6	0.7564
130	Shape transformation	11	0.6262
131	Size	8	0.6385
132	Smart device	8	1.0309
133	Smart textile	8	0.5689
134	Sme	5	0.7925
135	Smp	26	0.73
136	Smps	5	1.436
137	Soft actuator	8	0.8594
138	Soft robot	13	0.4577
139	Speed	14	0.967
140	Step	6	0.9574
141	Stereolithography	7	0.4933
142	Stiffness	9	0.9919
143	Stimuli responsive hydrogel	6	1.606
144	Stimuli responsive polymer	6	0.6044
145	Strain	11	0.6921
146	Strip	6	0.8447
147	Surface	12	0.449
148	Temporary shape	9	1.0233
149	Thermal stimulus	7	0.9333
150	Tissue	17	1.3708
151	Tissue engineering	16	0.878
152	Transition	10	0.5513
153	Variety	9	0.4681
154	Wide range	8	0.8176
155	Year	8	0.6544

1. Specimens without any stimulus.
2. Specimens with water stimulus (soaked at room temperature for 1 hour).
3. Specimens with heat stimulus (heating in an oven at temperature 50°C for 1 hour).
4. Specimens with hydrothermal stimulus (soaked in hot water (50°C) for 1 hour).

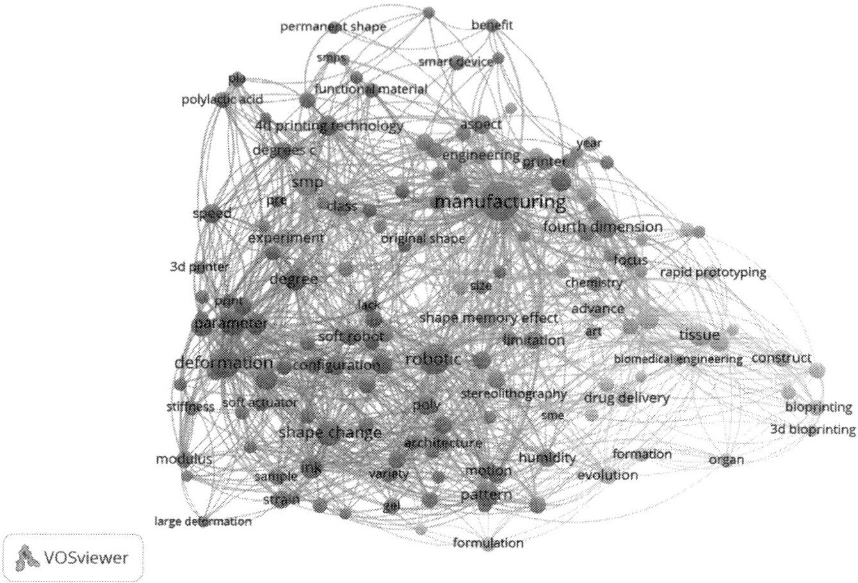

Figure 8.7 Networking diagram for keyword "stimulus in 4D printing."

8.2.4 Materials characterization

8.2.4.1 Melt flow index measurements

Melt flow index (MFI) test has been conducted to investigate rheological characteristics of the prepared samples. MFI is designated as a standard technique to scrutinize the rheological behavior for comparative purpose. In this method, flow of thermoplastics materials passes through an orifice for 10 minutes are measured. The test was conducted at a 40°C with applied weight (1 kg) as per ASTM D1238 method (Fig. 8.11). Standard test units for MFI are "g" in 10 minutes. The MFI measurements are taken for 3 times for each sample, and average value was recorded.

MFI values and viscosity calculated as per the procedure given by Shenoy et al. [24,25] of prepared samples are tabulated in Table 8.3.

8.2.4.2 Tensile testing

Tensile testing of the feedstock filaments has been carried out on universal tensile testing machine (UTM-SL-10A; Shanta Engineering, Mumbai). The specimens have a length of 200 mm (with 100 mm gage length and 50 mm for clamping on both sides) as shown in Fig. 8.12. After clamping in the

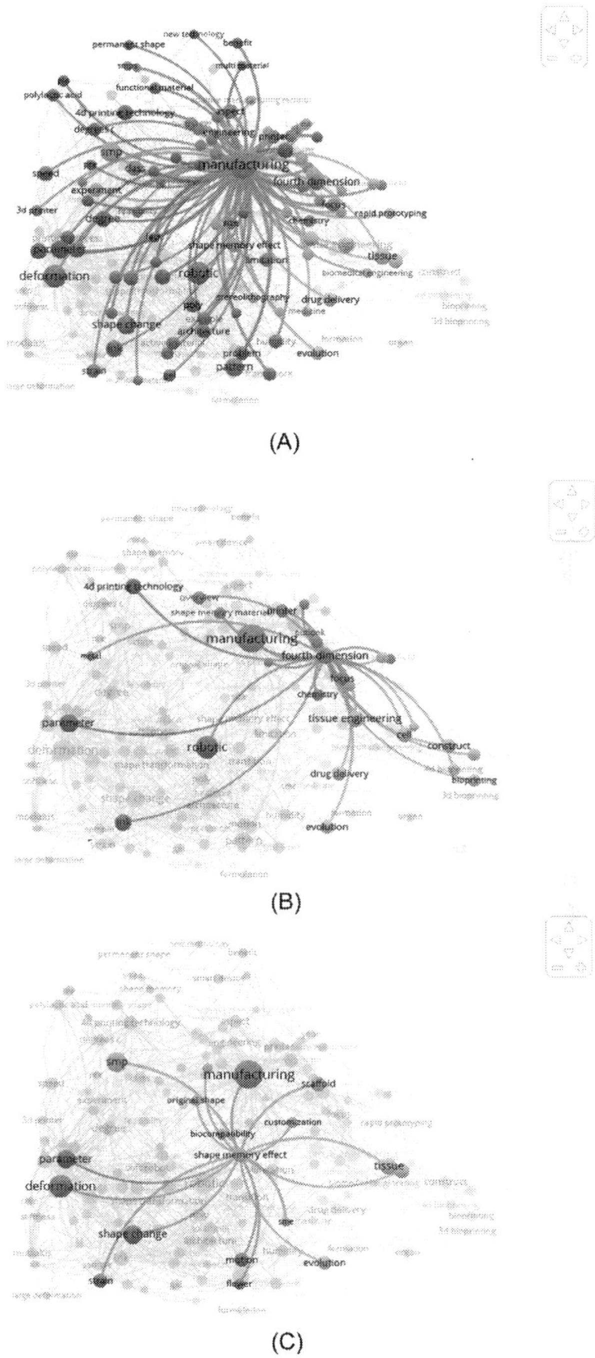

Figure 8.8 Research gap analysis by selecting nodal points as (A) manufacturing, (B) fourth dimension, and (C) shape memory.

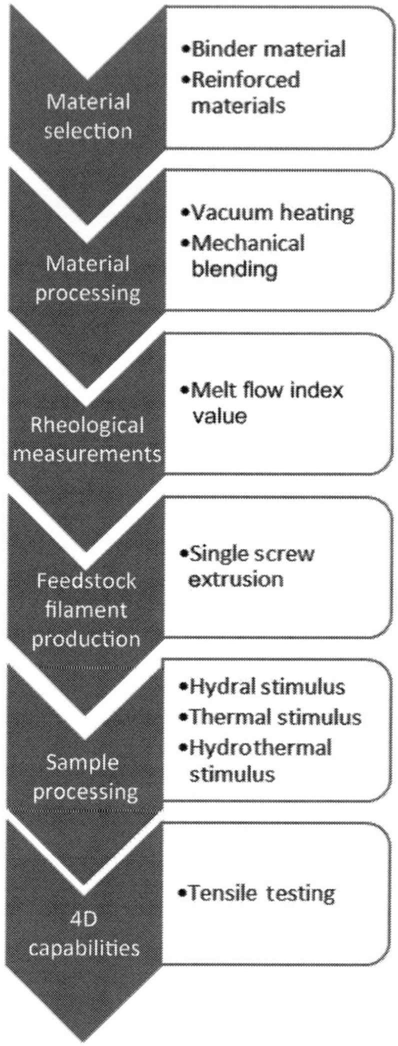

Figure 8.9 Process flow diagram.

Table 8.2 Weight proportions of the specimens.

S. No.	Composition	PA6	Al	Al_2O_3	Acrylonitrile butadiene styrene
1	A	60	26	14	–
2	B	60	28	12	–
3	C	60	30	10	–
4	ABS*	–	–	–	100

*Commercial feedstock filament P430.

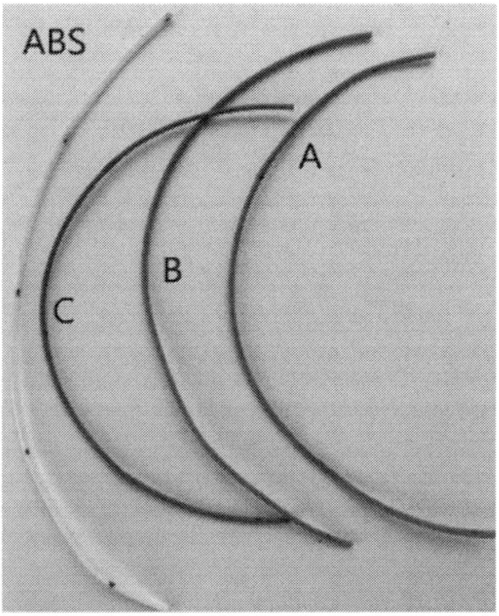

Figure 8.10 Specimens for hydrothermal testing.

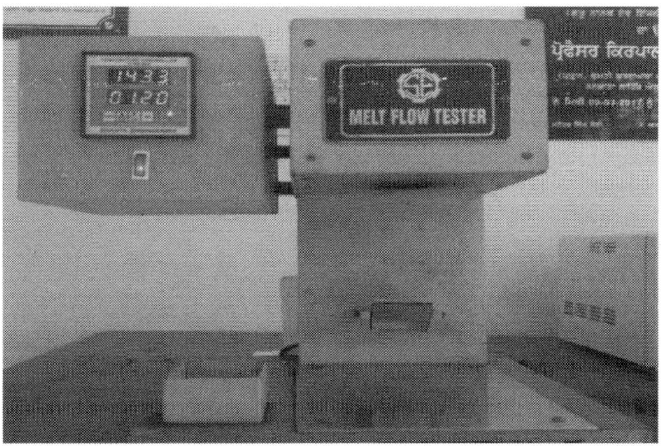

Figure 8.11 Melt flow index tester.

fixture, the sample was strained for 10 seconds by pulling at the rate of 30 mm/minute. By stretching the gage, length changes from 100 to 105 mm. Load was released after holding at a position of 105 mm for 5 seconds. The declamped specimens were place in room temperature under 1 atmospheric pressure for 2 hours. The % recovery in gage length (100 mm), and change of weight with water/heat stimulus is reported in Table 8.4.

Table 8.3 Melt flow index and viscosity values for different compositions/proportions (by weight %).

Composition	Density (ρ) g/cm^3	MFI g/(10 min)	Viscosity (μ) in (Pa-s)
A	1.52	2.19	13,158
B	1.51	2.25	12,722
C	1.50	*2.31*	12,310
Acrylonitrile butadiene styrene	1.05	*2.41*	8294

Note: The average value of three repeated observations has been quoted.

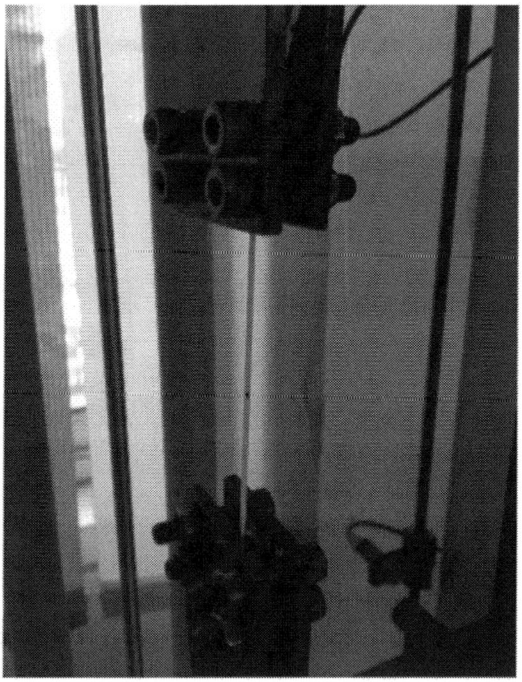

Figure 8.12 Tensile testing on universal testing machine.

8.3 Results and discussion

8.3.1 Rheological measurements

As shown in Fig. 8.13, the composition "C" has maximum MFI (2.31 g/10 minute) and A has minimum MFI (2.19 g/10 minute) values. The flow pattern of composition "C" was observed as similar to the virgin

Table 8.4 Effect of stimulus on % change in weight and % length recovery of specimens.

S. No.	Composition	Stimulus	Weight (g)		% Change in weight	Length recovery (%)
			Before	After		
1	A	No	0.623	0.623	0	83
2	B		0.615	0.615	0	86
3	C		0.607	0.607	0	90
4	Acrylonitrile butadiene styrene (ABS)		0.518	0.518	0	80
5	A	Water	0.623	0.673	8	82
6	B		0.615	0.655	6.5	84
7	C		0.607	0.632	4.1	88
8	ABS		0.518	0.523	0.9	80
9	A	Heat	0.623	0.620	− 0.4	89
10	B		0.615	0.613	− 0.3	92
11	C		0.607	0.606	− 0.1	96
12	ABS		0.518	0.515	− 0.5	85
13	A	Hydrothermal	0.623	0.673	8	85
14	B		0.615	0.655	6.5	86
15	C		0.607	0.632	2.6	92
16	ABS		0.518	0.518	0	83

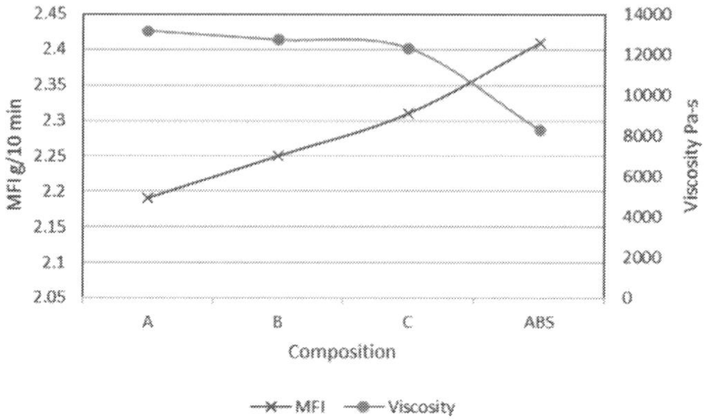

Figure 8.13 Melt flow index versus viscosity graph of compositions.

ABS, while operating under similar temperature and loading conditions. As a matter of fact, the MFI is inversely proportion to viscosity. Therefore as the MFI value decreases (due to the reinforcement), the viscosity increases. It is highlighted in Fig. 8.13 that there is an intersecting point of MFI curve and viscosity curve.

The benchmark MFI (MFI value of standard ABS material) is 2.41. The Al_2O_3 (due to its abrasive nature) causes the increase in shear stress in the matrix. Results indicated that, as the quantity of Al_2O_3 decrease, shear stress reduces which in turn decreases the value of fluid viscosity. The Al particles size is small (325 mesh) and has self-lubricating properties due to which the presence of Al reduces shear stress.

8.3.2 Tensile testing

Fig. 8.14 illustrates the % weight change and % length recovery of the specimens without any stimuli. There is no change in weight occurs during the testing but the % recovery (after placing the specimens for 2 hours under room temperature condition) has been varied among the specimens. The composition C has highest recovery (90%) followed by B (86%) and A (83%). The virgin ABS shows poor recovery (80%) in comparison with reinforced compositions. The reason behind this is due to the reinforcement in polymeric matrix, a small void remains in the composition. These voids are due to the different nature of polymeric and reinforced materials. As observed in the Fig. 8.14, the % recovery was also affected by the particle size. The amount of reinforcement is same in all the compositions but there is a variation of Al and Al_2O_3 proportions in

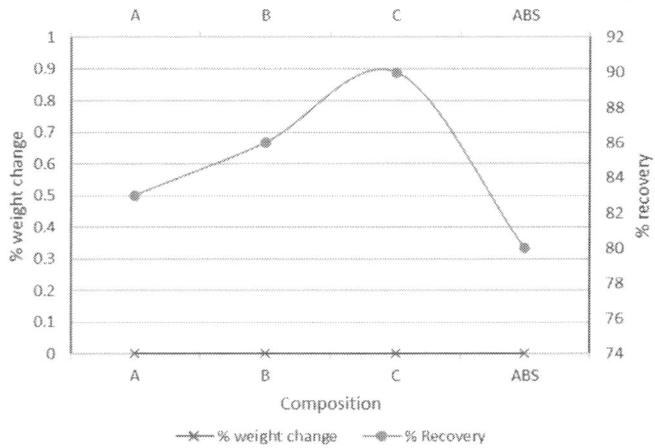

Figure 8.14 % weight change and % recovery of specimens without any stimulus.

the matrix. As already quoted, Al_2O_3has larger particle size (100 mesh) than Al (325 mesh). Larger particle size means large air gap at the interface of the reinforced particle and polymeric matrix. It has been also observed that, small air gap causes more % recovery. The reason is, large air gap causes necking during stretching. Therefore the composition C (which has large quantity of Al particles) has more recovery rate than other compositions and even ABS, which indicates its capability for 4D applications.

As the specimens exposed to water stimuli, the results show some changes, that is, water absorption among the specimens. Due to hydroscopic nature of PA6 (Nylon 6), the water absorption takes place as the samples experience water stimulus. Composition A shows large % of weight change followed by "B" and "C" (Fig. 8.15). The large water absorption in composition A is due to the large air gap in the composition. As we know, Al_2O_3 is amphoteric and insoluble in water. The virgin ABS also shows some changes in weight with water stimuli but it is very small (0.9%) in comparison with other reinforced samples.

On the other hand, the composition C has highest recovery rate. The recovery of composition A and B are nearly same but it is higher than ABS.

The heat stimulated specimens shows some different results of weight change and recovery rate. The loss of weight observed among all the specimens and it is higher for ABS sample. Among reinforced specimens, the composition A has large weight loss followed by B and C. As discussed already due to hydroscopic nature of polymeric materials, some moisture gets absorbed. As the specimens placed in the oven (50°C) for 1 hour, the water content evaporates, which causes loss of weight. Similar to previous results of tensile testing, the composition C has highest % recovery but in this case, it is higher (96%). A positive shift of % recovery observed among all the specimen (Fig. 8.16).

As mentioned already, the hydrothermal stimulus means, placing the specimen in a hot water (50°C) for 1 hour. Fig. 8.17 shows that the composition A shows higher changes in weight than other specimens. It should be noted that ABS shows no change in weight. On the other

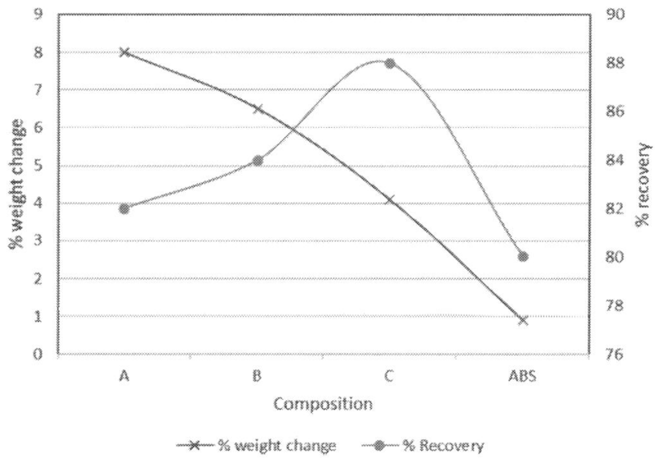

Figure 8.15 % weight change and % recovery of specimens with water stimulus.

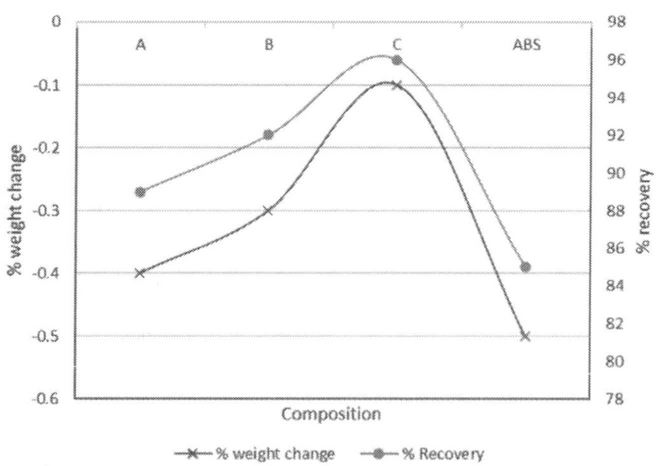

Figure 8.16 % weight change and % recovery of specimens with heat stimulus.

hand, after stretching the specimens shows % recovery in length. Similar to previous results, the composition C has highest % recovery (92%) and A has least (85%). ABS also shows 83% recovery in length.

As already specified above, there are four different test conditions considered for the experimentation, namely (1) specimens without any stimulus; (2) specimens with water stimulus (soaked at room temperature for 1 hour); (3) specimens with heat stimulus (heating in an oven at temperature 50°C for 1 hour); (4) specimens with hydrothermal stimulus (soaked in hot water (50°C) for 1 hour). Fig. 8.18 shows the % change in weight

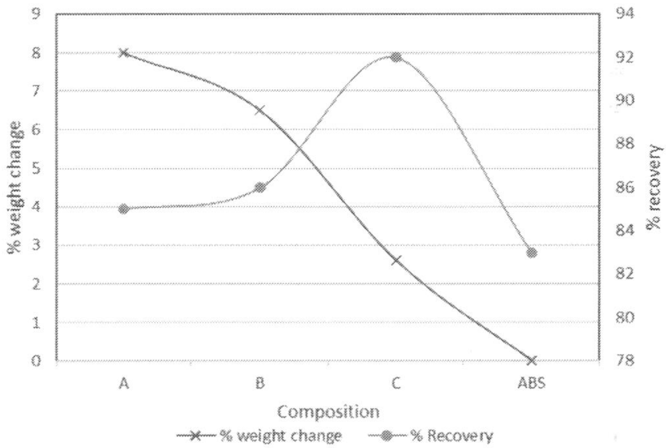

Figure 8.17 % weight change and % recovery of specimen with hydrothermal stimulus.

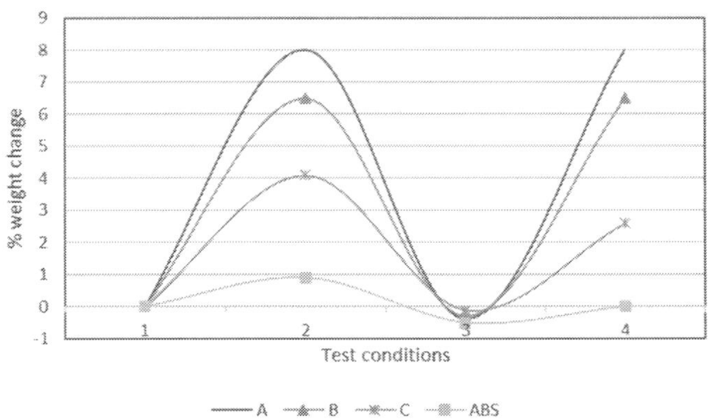

Figure 8.18 % weight change of specimens with different test conditions.

under the influence of different test conditions. The ABS material shows variation in weight change from −0.5% to 0.9%, and this the smallest range among all the specimens. In case of reinforced compositions, A varied in larger range (−0.4%–8%) followed by B (−0.3%–6.5%) and C (−0.1%–4.1%).

Similarly, Fig. 8.19 shows the % recovery of gage length (after releasing the tensile load and placing the specimens for 2 hours at room temperature) under the influence of different test conditions. All the specimens follow the same trend. The ABS material shows variation in from 80% to 85%, A from 82% to 89%, B from 84% to 92%, and C from 88% to 96%. It should be noted that composition C has higher length recovery rate and ABS has lowest.

8.3.3 Scanning electron microscopy

Particle morphology was assessed by using scanning electronic microscope (SEM) images (2D and 3D) as illustrated in Fig. 8.20A–C, respectively. All images have high magnification (at the scale of 500 μm). As observed in the images, Al and Al_2O_3 are available in PA6 matrix, but some agglomerates are also present. The reason of these agglomerates may be due to the cluster of small particles which indicate the inhomogeneity distribution of Al and Al_2O_3 particles in PA6 matrix. Although the care has been taken for the uniform particle size distribution in the matrix during the fabrication of feedstock filament, it can further be improved by reducing the particle size to nanoscale and selecting more suitable blending method (mechanical/chemical).

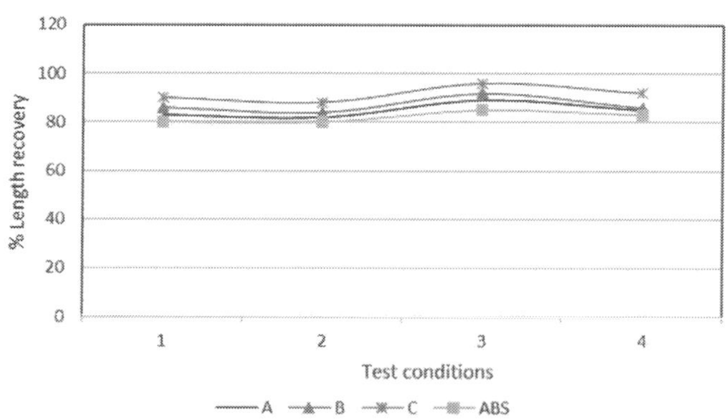

Figure 8.19 % recovery in length of specimens with different test conditions.

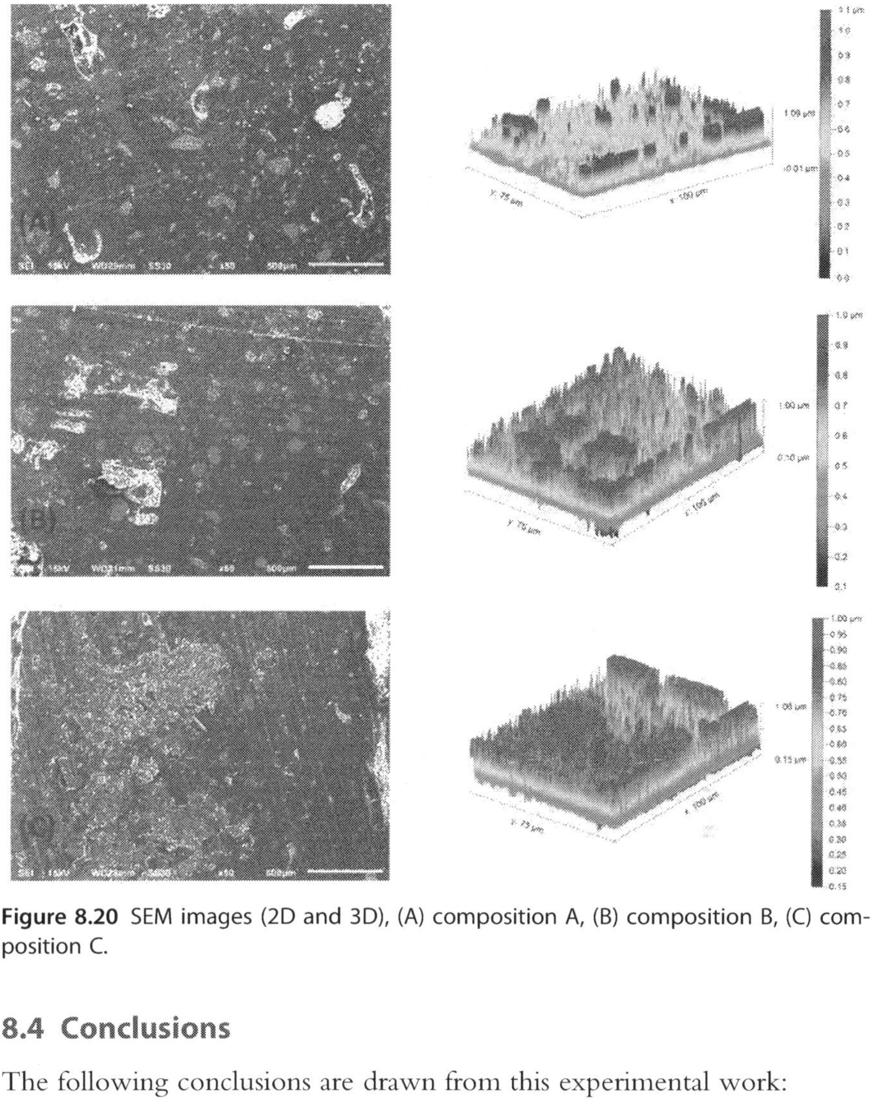

Figure 8.20 SEM images (2D and 3D), (A) composition A, (B) composition B, (C) composition C.

8.4 Conclusions

The following conclusions are drawn from this experimental work:

1. The suitability of alternative feedstock filament for FDM depends upon its flow behavior, which can be assessed by rheological measurements (MFI value). The composition "C" has maximum MFI (2.31 g/10 minute) and A has minimum MFI (2.19 g/10 minute) values. The flow pattern of composition "C" was observed as similar to the virgin ABS, while operating under similar temperature and loading conditions.
2. The ABS material shows variation in weight change from −0.5% to 0.9%, and this is the smallest range among all the specimens. In case of

reinforced compositions, composition A varied in larger range (− 0.4%–8%) followed by composition B (− 0.3%–6.5%) and C (− 0.1%–4.1%).

3. Under the influence of different test conditions, the ABS material shows variation in the % recovery of gage length from 80% to 85%, composition A from 82% to 89%, composition B from 84% to 92%, and composition C from 88% to 96%.
4. SEM images shows that agglomerates are present in the matrix, which may be due to the cluster of small particles. It shows the inhomogeneity distribution of Al and Al_2O_3 particles in PA6 matrix. Further work required for making improvement in particle size distribution.
5. The feedstock filament prepared by composition C has more length recovery rate after releasing the tensile load than other specimens, which indicates its capability for 4D applications.

Acknowledgement

The authors would like to acknowledge Manufacturing Research Lab, Guru Nanak Dev Engineering College, Ludhiana, India for providing their facilities to carry out present research work.

References

[1] Tibbits, S, The emergence of "4D printing." In: *Proceedings of the TED conferences*, 2013.
[2] Khoo ZX, Teoh JEM, Liu Y, Chua CK, Yang S, An J, et al. 3D printing of smart materials: a review on recent progresses in 4D printing. Virtual and Physical Prototyping 2015;10(3):103–22.
[3] Momeni F, Liu X, Ni J. A review of 4D printing. Materials & Design 2017;122:42–79.
[4] Kuang X, Roach DJ, Wu J, Hamel CM, Ding Z, Wang T, et al. Advances in 4D printing: materials and applications. Advanced Functional Materials 2019; 29(2):1805290.
[5] Sun L, Huang WM, Ding Z, Zhao Y, Wang CC, Purnawali H, et al. Stimulus-responsive shape memory materials: a review. Materials & Design 2012;33:577–640.
[6] Zhou Y, Huang WM, Kang SF, Wu XL, Lu HB, Fu J, et al. From 3D to 4D printing: approaches and typical applications. Journal of Mechanical Science and Technology 2015;29:4281–8.
[7] Zhou J, Sheiko SS. Reversible shape-shifting in polymeric materials. Journal of Polymer Science Part B: Polymer Physics 2016;54:1365–80.
[8] Yu K, Ge Q, Qi HJ. Reduced time as a unified parameter determining fixity and free recovery of shape memory polymers. Nature Communications 2014;5(1):1–9.
[9] Ge Q, Luo X, Iversen CB, Mather PT, Dunn ML, Qi HJ. Mechanisms of triple-shape polymeric composites due to dual thermal transitions. Soft Matter 2013; 9(7):2212–23.

[10] Meier, H., et al. Selective laser melting of NiTi shape memory components. In: Advanced research in virtual and rapid prototyping, Leiria, Portugal. 2009.
[11] Meier, H, Haberland, C, Frenzel, J,. Structural and functional properties of NiTi shape memory alloys produced by selective laser melting. In: London: innovative developments in virtual and physical prototyping, 2012, p. 291–96.
[12] Dadbakhsh S, et al. Effect of SLM parameters on transformation temperatures of shape memory nickel titanium parts. Advanced Engineering Materials 2014;16:1140–6.
[13] Uchino K. The development of piezoelectric materials and the new perspective. In: Uchino K, editor. Advanced piezoelectric materials—science and technology. Padstow, Cornwall: Woodhead Publishing; 2010. p. 1–43.
[14] Kim K, et al. 3D optical printing of piezoelectric nanoparticle- polymer composite materials. ACS Nano 2014;8:9799–806.
[15] Lin D, et al. Three-dimensional printing of complex structures: man-made or toward nature? ACS Nano 2014;8:9710–15.
[16] Tibbits S. 4D printing: multi-material shape change. Architectural Design 2014; 84(1):116–21.
[17] Raviv D, Zhao W, McKnelly C, Papadopoulou A, Kadambi A, Shi B, et al. Active printed materials for complex self-evolving deformations. Scientific Reports 2014;4:7422.
[18] Jamal M, Kadam SS, Xiao R, Jivan F, Onn TM, Fernandes R, et al. Bio-Origami hydrogel scaffolds composed of photocrosslinked PEG bilayers. Advanced Healthcare Materials 2013;2:1142–50.
[19] Lewis JA. Direct ink writing of 3D functional materials. Advanced Functional Materials 2006;16:2193–204.
[20] Lebel LL, Aissa B, Khakani MAE, Therriault D. Ultraviolet-assisted direct-write fabrication of carbon nanotube/polymer nanocomposite microcoils. Advanced Materials 2010;22:592–6.
[21] Gratson GM, Lewis JA. Phase behavior and rheological properties of polyelectrolyte inks for direct-write assembly. Langmuir 2005;21:457–64.
[22] Singh Boparai K, Singh R, Singh H. Experimental investigations for development of Nylon6-Al-Al_2O_3 alternative FDM filament. Rapid Prototyping Journal 2016; 22(2):217–24.
[23] Boparai KS, Singh R, Singh H. Modeling and optimization of extrusion process parameters for the development of Nylon6–Al–Al_2O_3 alternative FDM filament. Progress in Additive Manufacturing 2016;1:115–28.
[24] Shenoy AV, Saini DR, Nadkarni VM. Rheology of nylon 6 containing metal halides. Journal of Materials Science 1983;18:2149–55.
[25] Shenoy AV, Saini DR, Nadkarni VM. Melt rheology of polymer blends from melt flow index. International Journal of Polymeric Materials 1984;10(3):213–35.

CHAPTER 9

On PLA–ZnO composite matrix for shape memory effect

Ranvijay Kumar[1], Rupinder Singh[2], Vinay Kumar[3] and Pawan Kumar[4]
[1]Department of Mechanical Engineering, University Center for Research and Development, Chandigarh University, Mohali, India
[2]Department of Mechanical Engineering, National Institute of Technical Teachers Training and Research, Chandigarh, India
[3]Department of Production Engineering, Guru Nanak Dev Engineering College, Ludhiana, India
[4]Department of Physics, University Institute of Science, Chandigarh University, Mohali, India

9.1 Introduction

The shape memory polymers are the essential component for the manufacturing of smart actuators and sensors. The application of the smart polymer-based structures includes wrinkle-free fabrics, robotics, self-assembly, assembly automation, autochoke element for engines, photonics, ophthalmic devices includes punctual and intraocular lenses, preparation of vascular stents, surgical sutures, automotive fenders, temper-evident seals, etc. Polylactic acid (PLA) has been considered as one of the shape memory polymers which has good biocompatibility as well as biodegradability, and it is responsive to heat. Senatov et al. have investigated the shape memory behavior of 3D printed PLA-hydroxiapatite-based porous structures. The shape recovery has been reported up to 98% in this case [1]. The composites of polyurethane (PU)-PLA-carbon nanotubes have also shown the shape memory effect [2]. The biodegradable polycaprolactone (PCL)-PLA blend has shown the shape memory behavior [3]. The study has been reported for increase in the shape recovery by the addition of thermoplastic polyeurathene (TPU) in PLA matrix. Also, increasing the predeformation temperature has improved the shape memory behavior [4]. The surprising results have been obtained for shape memory effect when toughening of PLA made by polyamide elastomer [5]. The developments of temperature sensitive shape memory nanocomposites have also secured the applications in sensors and actuators. The nanocomposite prepared by incorporation of modified grapheme nanoplatelets has been reported with shape memory behavior [6]. Similarly, the shape memory has been also reported for the

4D Printing
DOI: https://doi.org/10.1016/B978-0-12-823725-0.00008-4

combination of electrospun PLA nanofiners plasticized with oligomer [7], cellulose nanocrystal addition in PLA-epoxidized natural rubber (ENR) [8], PLA filaments [9], PLA-ENR-Fe_3O_4 thermoplastic valcanizates (TPVs) [10], PLA-natural rubber (NV)-SiO_2 TPVs [11], and PLA-TPU blends [12,13].

Zinc oxide (ZnO) is a valuable water insoluble semiconductor materials which when added to other materials modify the behavior. The ZnO has been used as an additive in numerous materials such as cosmetics, food supplements, rubbers, polymers, metals, glass, batteries, paints, etc. for improving the functionalities. ZnO also modifies the shape memory behavior when reinforced with polymeric materials. Gao et al. have studied the superelasticity and nanofracture behavior of ZnO nanohelices which is a shape memory ceramic nanostructure [14]. The study has been conducted for shape memory behavior of PU-PCL-ZnO ternary blends. The study has revealed that shape memory effect can be increased with optimum concentration of ZnO (20 wt.%) in PU-PCL blend [15]. Similarly, the addition of ZnO nanostructures has been reported for the induced shape memory effect when reinforced with starch/lysine [16], NiTi shape memory alloys [17], PLA [18], TPU [19], ethylene-propylene-diene rubber/polypropylene (EPDM/PP) TPVs [20], PU [21], stretched poly(methyl methacrylate) (PMMA) [22], etc.

To meet the previous investigations of shape memory effect of PLA polymer, the database available on http://www.webofknowledge.com has been taken for analysis. The target year has been selected between the duration of 1999 and 2021. Total of 118 studies have been explored till year 2121 by searching with keyword "PLA and shape memory effect." The data have been analyzed on VOS viewer software package. On analyzing the database, it has been retrieved a total of 12032 terms and by filtering the 10 minimum number of occurrences of terms, 61 terms have met the threshold of the study. Further, the desirable 29 terms have been selected for the analysis. Table 9.1 shows the selected term, their occurrences, and relevance score.

Fig. 9.1 shows the bibliographic map-based network for keyword "PLA and shape memory effect." It has been observed that there are four different clusters are formed. The previous studies have been reported for the investigations PLA polymer in relation to tensile strength, crystallization, applications, fabrication, shape recovery, shape memory effect, morphology, dynamic mechanical analysis, nanocomposites, differential scanning calorimetry, deformation mechanism, and blending with PCL, TPU, etc.

In relation to shape memory investigation of PLA polymer, previous studies have been reported mostly for the development of materials,

Table 9.1 Selected term, their occurrences, and relevance score.

Sr. No.	Term	Occurrences	Relevance score
1	Shape recovery ratio	15	1.803
2	Tensile strength	22	1.713
3	Elongation	13	1.6686
4	Fabrication	15	1.6528
5	PCL	10	1.5995
6	Polylactide	25	1.5349
7	Thermoplastic polyurethane	14	1.2681
8	Nanocomposite	16	1.1784
9	Crystallinity	15	1.1746
10	Differential scanning calorimetry	16	1.1518
11	Dynamic mechanical analysis	11	1.1468
12	Shape memory effect	54	1.018
13	Incorporation	14	0.9648
14	Application	36	0.948
15	Crystallization	21	0.9199
16	Polylactic acid	35	0.8464
17	Shape memory polymer	36	0.8154
18	Shape memory behavior	25	0.7937
19	Deformation	16	0.7775
20	Polymer	33	0.743
21	Morphology	22	0.6501
22	Shape recovery	19	0.6347
23	Characterization	16	0.6267
24	Addition	26	0.6251
25	Blend	41	0.605
26	Glass transition temperature	20	0.5493
27	Temperature	52	0.5468
28	Mechanical property	41	0.5354
29	Development	11	0.5085

Source: From http://www.webofknowledge.com.

prototypes, shape memory behavior, and taking temperature as the stimulus. It needs more to explore the shape memory PLA with effect and investigations of crystallinity, differential scanning calorimetry, nanocomposites, morphology, dynamic mechanical analysis, deformation mechanism, glass transition temperature, etc. (Fig. 9.2).

From the literature survey, it is evident that number of studies has been reported on investigation of shape memory behavior of the PLA polymers. Some studies have been reported on investigation of temperature as the stimulus for shape memory effect. Also, previous studies have outlined that

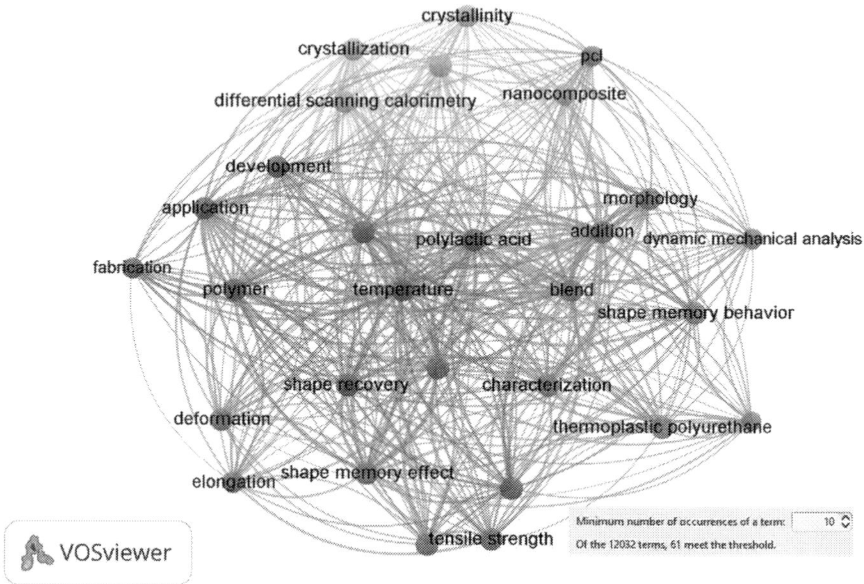

Figure 9.1 Bibliographic map-based network for keyword "PLA and shape memory effect" (color of the nodes represent different cluster formation). *From http://www.webofknowledge.com.*

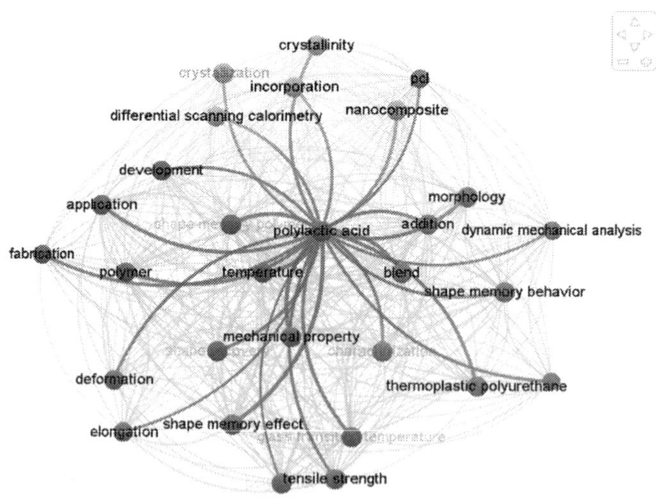

Figure 9.2 Bibliographic network analysis for gaps in shape memory effect of PLA (colors of the nodes represent different clusters).

ZnO reinforcement in polymers exhibits shape memory behavior. But hitherto, little has been reported on investigating the shape memory effect by varying the heat treatment conditions. In this study, the effect of heat treatment, processing temperature, and screw torque on the shape memory effect of PLA–ZnO composites feedstock filaments has been explored.

9.2 Materials and methods

Fig. 9.3 shows the detailed methodology for the PLA–ZnO composite feedstock filament preparations and shape memory investigations. The PLA in granules form (diameter: 1–2 mm) has been purchased from Batra Polymer Pvt Ltd. (Ludhiana, India). The ZnO nanoparticles have been

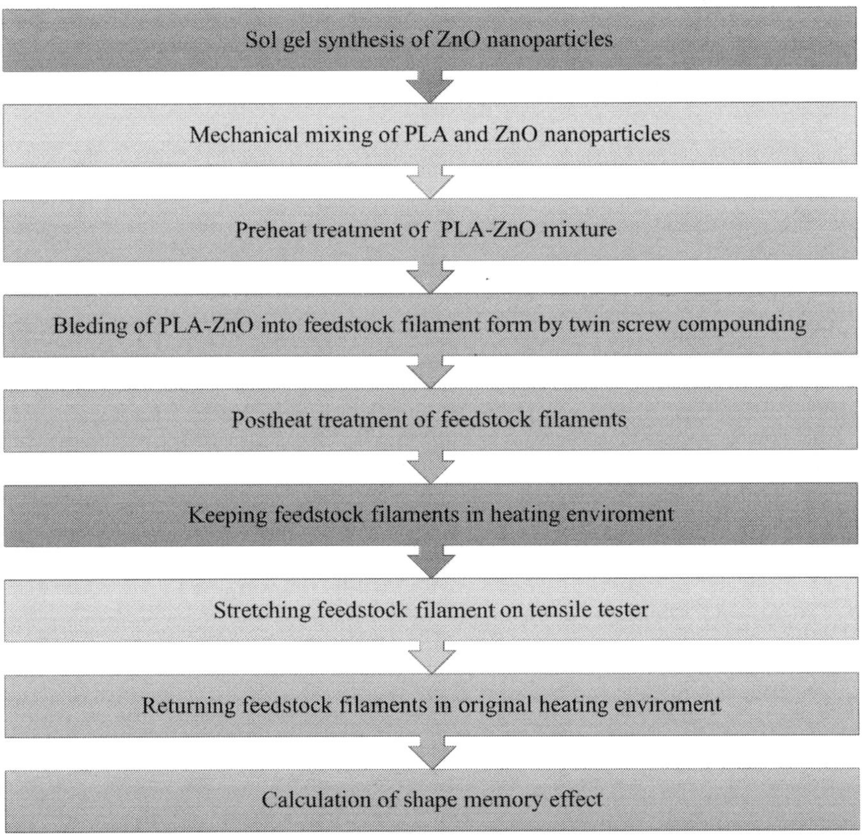

Figure 9.3 Methodology for PLA–ZnO composite filament preparation and shape memory investigations.

synthesized by well-established sol-method. In next step, the ZnO (2 wt.%) nanoparticles have been mechanically mixed with PLA polymer granules in the ratio of 98 (PLA)−2% (ZnO) (wt.%) with the help of linseed oil.

The preheat treatment of mixture has been performed in the oven at 60°C (near glass transition temperature of PLA) for 1 hour. The blending of PLA−ZnO by varying processing temperature and screw torque has been performed for preparation of feedstock filament. The twin screw compounder has been used to process the feedstock filaments. The prepared feedstock filaments were put in oven at 60°C for 1 hour as postmanufacturing heat treatment. Further, the feedstock filaments have been stretched on tensile tester and put in original heating condition to memorize their shape. The results have been observed for the shape memory effect of PLA−ZnO composite feedstock filaments.

9.3 Experimentation

9.3.1 Twin screw compounding

The mixture of PLA and ZnO nanoparticles was blended on twin screw compounder (Make: Mettler Toledo; Maximum temperature: 300°C). Fig. 9.4 shows the pictorial view of twin screw compounder and prepared feedstock filament.

Table 9.2 shows the design of experiment for PLA−ZnO feedstock filament preparation based upon Taguchi L9 orthogonal array. Three different levels of the each parameter have been selected. The processing condition has been selected as not-treated samples, preheat treated samples of filaments, and postheat treated heat treated filaments. The filament processing temperature has been varied to 170°C, 180°C, and 190°C, and screw temperature has been varied to 0.10, 0.20, and 0.30 N m for feedstock filament manufacturing. It should be noted that the levels of the process parameters such as filament processing temperature and screw torque have been selected based upon uniformity of the manufactured feedstock filaments. Beyond this range of process parameters, the filaments were observed dimensionally distorted.

9.3.2 Shape memory investigation

The shape memory behavior of the PLA−ZnO feedstock filaments has been investigated on effect of taking temperature as the external stimulus. The manufactured feedstock filaments were first put in oven at 60°C for

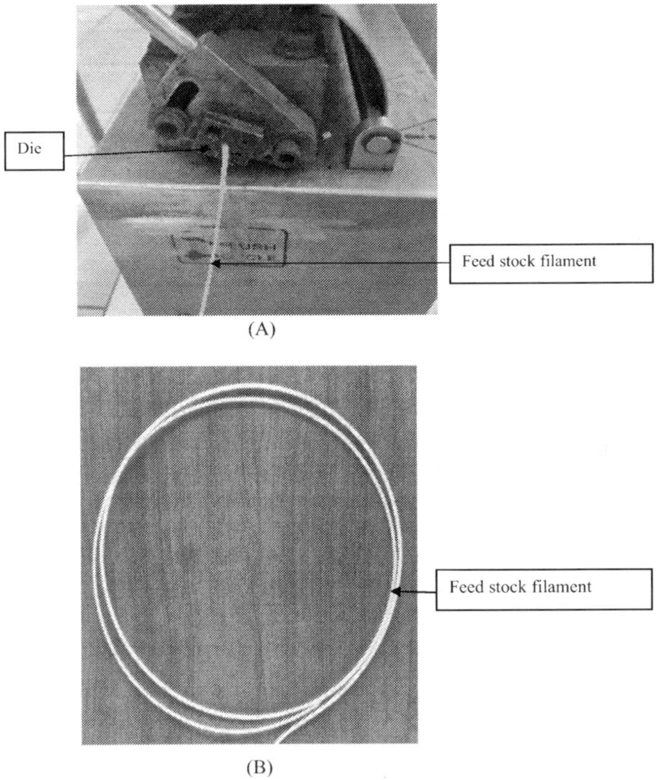

Figure 9.4 Twin screw compounder (A), prepared feedstock filament (B).

Table 9.2 Experimental design for manufacturing of feedstock filaments.

S. No.	Processing condition	Filament processing temperature (°C)	Screw torque (N m)
1	Nontreated	170	0.10
2	Nontreated	180	0.20
3	Nontreated	190	0.30
4	Pretreated	170	0.20
5	Pretreated	180	0.30
6	Pretreated	190	0.10
7	Posttreated	170	0.30
8	Posttreated	180	0.10
9	Posttreated	190	0.20

1 hour and then stained on the tensile tester (Make: Shanta Engineering; Maximum capacity: 5000 N) with the stain rate of 20 mm/min for 6 seconds. The dimensions before stretching and after stretching were measured. Again the stretched filaments were kept in original temperature condition (60°C for 1 hour) to memorize its shape. Further final dimension of the recovered feedstock filaments has been measured with calculation of shape memory effect.

9.4 Results and discussion

Table 9.3 shows the information of shape memory behavior of the manufactured feedstock filament samples. The experiments have been repeated three times for each sample and results are given in form of mean value. It has observed the minimum shape recovery in case of sample 1 (manufactured without heat treatment at 170°C temperature and 0.10 N m torque) and maximum in case of sample 9 (manufactured with postheat treatment at 190°C processing temperature and 0.20 N m torque). The percentage shape recovery has been observed 23.68% and 60.68% for sample 1 and sample 9, respectively. The better shape memory in case of sample 9 has been observed. This may due to the postheat treatment, and higher temperature processing condition. The postheat treatment has played an important role in the manufacturing of less defected filaments and higher processing temperature for better dispersion of the ZnO nanoparticles in PLA matrix.

Table 9.3 Shape memory behavior of PLA–ZnO feedstock filaments.

S. No.	Processing condition	Initial length (mm)	Strain produce (mm)	Total length after strain (mm)	Final length after recovery (mm)	Percentage shape recovery (%)
1	Nontreated	93.63	1.52	95.15	94.79	23.68
2	Nontreated	95.75	1.58	97.33	96.85	30.38
3	Nontreated	94.73	1.14	95.87	95.48	34.21
4	Pretreated	91.66	1.57	93.23	92.70	33.75
5	Pretreated	97.39	1.55	98.94	98.18	49.03
6	Pretreated	94.8	1.56	96.36	95.63	46.79
7	Posttreated	92.82	1.54	94.36	93.73	40.91
8	Posttreated	95.84	1.07	96.91	96.30	57.01
9	Posttreated	89.57	1.54	91.11	90.18	60.39

Table 9.4 SN ratio for percentage shape recovery of PLA–ZnO filaments.

S. No.	Processing condition	SN ratio for percentage shape recovery (dB)
1	Nontreated	27.48
2	Nontreated	29.65
3	Nontreated	30.68
4	Pretreated	30.56
5	Pretreated	33.81
6	Pretreated	33.40
7	Posttreated	32.23
8	Posttreated	35.12
9	Posttreated	35.62

The better filament processing has been led to the providing the better shape memory characteristics. On the other hand, the filament manufactured without any heat treatment (either preheat treatment or postheat treatment), lower processing temperature, and lower screw torque has exhibited low shape recovery in case of sample 1. The lower shape recovery in case of sample 1 is may be due to the fact that inadequate processing has increased the manufacturing defect which is responsible minimum recovery.

To find the optimum process parameters for maximizing the shape recovery and predicted percentage shape recovery, the signal-to-noise (SN) ratio has been calculated for each experimental condition by considering "Larger is better" type case. Table 9.4 shows the SN ratio for percentage shape recovery of PLA–ZnO composite filaments.

Fig. 9.5 shows the main effect plot for SN ratio of percentage shape recovery of PLA–ZnO composite feedstock filaments. As per the calculation of SN ratio, it has been predicted a set of process parameters which can maximize the SN ratio. The predicted setting of feedstock filament manufacturing has been obtained as the combination of postheat treatment, 190°C processing temperature and 0.30 N m screw torque.

The analysis of variance (ANOVA)-based statistical analysis has been perfumed to find the value of percentage shape recovery at predicted manufacturing setting. Table 9.5 shows the ANOVA for SN ratio of percentage shape recovery of PLA–ZnO filaments. The percentage contributions of heat treatment, processing temperature, and screw torque have been obtained 68.43%, 30.51%, and 0.24%, respectively. The percentage residual error has been obtained 0.78% which is below 5% shows that process commands strong hold over varying process parameters. The

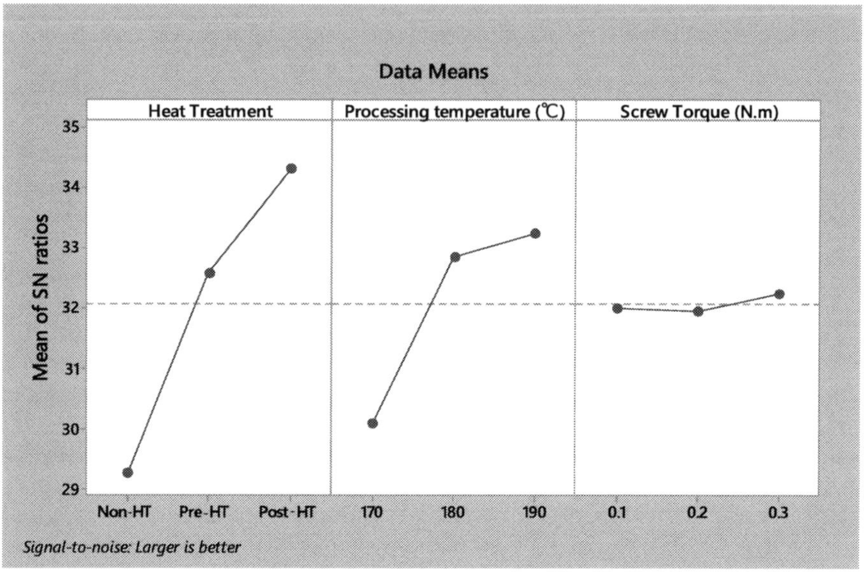

Figure 9.5 Main effect plot for SN ratio of percentage shape recovery of PLA–ZnO composite feedstock filaments.

Table 9.5 ANOVA for SN ratio of percentage shape recovery of PLA–ZnO filaments.

Source	DF	Seq SS	Adj SS	Adj MS	*F*	*P*	Percentage contribution
Heat treatment	2	39.52	39.52	19.76	86.78	0.011	68.43
Processing temperature (°C)	2	17.62	17.62	8.81	38.71	0.025	30.51
Screw torque (N m)	2	0.14	0.14	0.07	0.33	0.753	0.24
Residual error	2	0.45	0.45	0.23			0.78
Total	8	57.75					

DF, Degree of freedom; *F*, Fishers value; *Ms*, mean of squares; *P*, probability; *SS*, sum of squares.

process capability has been also verified by the controlled *P* value for heat treatment ($P = .011$) and processing temperature ($P = .025$) which obtained lesser than 0.05.

Table 9.6 shows the response of SN ratio for percentage shape recovery of PLA–ZnO composite feedstock filaments. As per the variation over SN ratios, the heat treatment has ranked 1, screw torques ranked 3, and processing temperature ranked 3.

Table 9.6 Response table for SN ratio of percentage shape recovery of PLA–ZnO composite feedstock filaments.

Level	Heat treatment	Processing temperature (°C)	Screw torque (N m)
1	29.27	30.10	32.00
2	32.59	32.86	31.95
3	34.32	33.24	32.24
Delta	5.05	3.14	0.30
Rank	1	2	3

The value of percentage shape recovery has been calculated on predicted manufacturing setting. As per the prediction given in Fig. 9.5 for SN ratios, the predicted setting has been obtained for maximizing percentage recovery with the manufacturing combination of postheat treatment, 190°C processing temperature, and 0.3 N m screw toques. The value of percentage shape recovery (y_{opt}) at predicted setting can be calculated by given formula:

$$y^2{}_{opt} = (10)^{\eta_{opt}/10}$$

where η_{opt} is the optimum SN ratio of percentage shape recovery. The η_{opt} can be calculated as:

$$\eta_{opt} = m + (m_{a3} - m) + (m_{b3} - m) + (m_{c3} - m)$$

where m is the mean of SN ratio of 9 different set of experiment (Table 9.4), m_{a3} is the AN ratio of heat treatment at level 3, m_{b3} is the SN ratio processing temperature a level 3, and m_{c3} is the SN ratio of screw torque at level 3.

The values are obtained from Tables 9.4 and 9.6. Putting value to find out η_{opt}

$$\eta_{opt} = 32.06 + (34.32 - 32.06) + (33.24 - 32.06) + (32.24 - 32.06)$$

$$\eta_{opt} = 35.68\text{dB}$$

where the value of y_{opt} has been calculated as:

$$y^2{}_{opt} = (10)^{\eta_{opt}/10}$$

$$y^2{}_{opt} = (10)^{35.68/10}$$

$$y_{opt} = 60.81\%$$

The value of percentage shape memory has been obtained 60.81% for the predicted set of process parameters. The percentage shape memory has been obtained 61.08% which is very closer to the predicted value.

9.5 Summary

- The percentage shape recovery has been observed 23.68% and 60.68% for sample 1 and sample 9, respectively. The better shape memory in case of sample 9 has been observed may be due to the fact that the action of postheat treatment and higher thermal processing condition.
- The postheat treatment has played an important role in the manufacturing of less defected filaments and higher processing temperature for better dispersion of the ZnO nanoparticles in PLA matrix.
- The predicted setting of feedstock filament manufacturing has been obtained as the combination of postheat treatment, 190°C processing temperature, and 0.30 N m screw torque.
- The value of percentage shape memory has been obtained as 60.81% for the predicted set of process parameters. The percentage shape memory has been obtained 61.08% which is very closer to the predicted value.

Acknowledgments

The authors are highly thankful to University Centre for Research and Development, Chandigarh University and Centre for Manufacturing Research, Guru Nanak Dev Engineering College, Ludhiana for technical assistance.

References

[1] Senatov FS, Niaza KV, Zadorozhnyy MY, Maksimkin AV, Kaloshkin SD, Estrin YZ. Mechanical properties and shape memory effect of 3D-printed PLA-based porous scaffolds. Journal of Mechanical Behavior of Biomedical Materials 2016;57:139–48.

[2] Raja M, Ryu SH, Shanmugharaj AM. Thermal, mechanical and electroactive shape memory properties of polyurethane (PU)/poly(lactic acid) (PLA)/CNT nanocomposites. European Polymer Journal 2013;49(11):3492–500.

[3] Navarro-Baena I, Sessini V, Dominici F, Torre L, Kenny JM, Peponi L. Design of biodegradable blends based on PLA and PCL: from morphological, thermal and mechanical studies to shape memory behavior. Polymer Degradation and Stability 2016;132:97–108.

[4] Lai SM, Lan YC. Shape memory properties of melt-blended polylactic acid (PLA)/thermoplastic polyurethane (TPU) bio-based blends. Journal of Polymer Research 2013;20(5):1–8.

[5] Zhang W, Chen L, Zhang Y. Surprising shape-memory effect of polylactide resulted from toughening by polyamide elastomer. Polymer 2009;50(5):1311–15.
[6] Keramati M, Ghasemi I, Karrabi M, Azizi H, Sabzi M. Incorporation of surface modified graphene nanoplatelets for development of shape memory PLA nanocomposite. Fibers and Polymers 2016;17(7):1062–8.
[7] Leones A, Sonseca A, López D, Fiori S, Peponi L. Shape memory effect on electrospun PLA-based fibers tailoring their thermal response. European Polymer Journal 2019;117:217–26.
[8] Cao L, Liu C, Zou D, Zhang S, Chen Y. Using cellulose nanocrystals as sustainable additive to enhance mechanical and shape memory properties of PLA/ENR thermoplastic vulcanizates. Carbohydrate Polymers 2020;230:115618.
[9] Radjabian M, Kish MH, Mohammadi N. Structure–property relationship for poly (lactic acid) (PLA) filaments: physical, thermomechanical and shape memory characterization. Journal of Polymer Research 2012;19(6). Available from: https://doi.org/10.1007/s10965-012-9870-0.
[10] Huang J, Cao L, Yuan D, Chen Y. Design of multi-stimuli-responsive shape memory biobased PLA/ENR/Fe3O4 TPVs with balanced stiffness–toughness based on selective distribution of Fe3O4. ACS Sustainable Chemistry & Engineering 2018;7(2):2304–15.
[11] Liu Y, Huang J, Zhou J, Wang Y, Cao L, Chen Y. Influence of selective distribution of SiO2 nanoparticles on shape memory behavior of co-continuous PLA/NR/SiO2 TPVs. Materials Chemistry Physics 2020;242:122538.
[12] Dogan SK, Boyacioglu S, Kodal M, Gokce O, Ozkoc G. Thermally induced shape memory behavior, enzymatic degradation and biocompatibility of PLA/TPU blends: "effects of compatibilization". Journal of Mechanical Behavior of Biomedical Materials 2017;71:349–61.
[13] Ahmed MF, Li Y, Yao Z, Cao K, Zeng C. TPU/PLA blend foams: enhanced foamability, structural stability, and implications for shape memory foams. Journal of Applied Polymer Science 2019;136(17):47416.
[14] Gao PX, Mai W, Wang ZL. Superelasticity and nanofracture mechanics of ZnO nanohelices. Nano Letters 2006;6(11):2536–43.
[15] Abbasi-Shirsavar M, Baghani M, Taghavimehr M, Golzar M, Nikzad M, Ansari M, et al. An experimental–numerical study on shape memory behavior of PU/PCL/ZnO ternary blend. Journal of Intelligent Material Systems and Structures 2019;30(1):116–26.
[16] Kotharangannagari VK, Krishnan K. Biodegradable hybrid nanocomposites of starch/lysine and ZnO nanoparticles with shape memory properties. Materials & Design 2016;109:590–5.
[17] Sabzi M, Far SM, Dezfuli SM. Characterization of bioactivity behavior and corrosion responses of hydroxyapatite-ZnO nanostructured coating deposited on NiTi shape memory alloy. Ceramics International 2018;44(17):21395–405.
[18] Kumar R, Singh R, Singh M, Kumar P. ZnO nanoparticle-grafted PLA thermoplastic composites for 3D printing applications: tuning of thermal, mechanical, morphological and shape memory effect. Journal of Thermoplastic Composite Materials 2020. Available from: https://doi.org/10.1177/0892705720925119.
[19] Koerner H, Kelley J, George J, Drummy L, Mirau P, Bell NS, et al. ZnO nanorod–thermoplastic polyurethane nanocomposites: morphology and shape memory performance. Macromolecules. 2009;42(22):8933–42.
[20] Xu C, Wu W, Zheng Z, Wang Z, Nie J. Design of shape-memory materials based on sea-island structured EPDM/PP TPVs via in-situ compatibilization of methacrylic acid and excess zinc oxide nanoparticles. Composites Science and Technology 2018;167:431–9.

[21] Wang Y, Zhang P, Zhao Y, Dai R, Huang M, Liu W, et al. Shape memory composites composed of polyurethane/ZnO nanoparticles as potential smart biomaterials. Polymer Composites 2020;41(5):2094–107.
[22] Lu H, Liu Y, Huang WM, Wang C, Hui D, Fu YQ. Controlled evolution of surface patterns for ZnO coated on stretched PMMA upon thermal and solvent treatments. Composites Part B: Engineering 2018;132:1–9.

Index

Note: Page numbers followed by "*f*" and "*t*" refer to figures and tables, respectively.

M

N

O

P

U

V

W

Z

FSC
www.fsc.org
MIX
Papier | Fördert gute Waldnutzung
FSC® C083411

Kolibri 360 GmbH
Ferdinand-Jühlke-Straße 7
99095 Erfurt, Deutschland
produktsicherheit@kolibri360.de